The Dynamics of Energy

Supply, Conversion, and Utilization

The Dynamics of Energy

Supply, Conversion, and Utilization

Horacio Perez-Blanco

CRC Press
Taylor & Francis Group
Boca Raton London New York

CRC Press is an imprint of the
Taylor & Francis Group, an **informa** business

CRC Press
Taylor & Francis Group
6000 Broken Sound Parkway NW, Suite 300
Boca Raton, FL 33487-2742

First issued in paperback 2017

© 2009 by Taylor and Francis Group, LLC
CRC Press is an imprint of Taylor & Francis Group, an Informa business

No claim to original U.S. Government works

ISBN 13: 978-1-138-11273-5 (pbk)
ISBN 13: 978-1-4200-7688-2 (hbk)

**Visit the Taylor & Francis Web site at
http://www.taylorandfrancis.com**

**and the CRC Press Web site at
http://www.crcpress.com**

Contents

A Software Notice

This book is designed to develop competence in energy matters on three levels: conceptual, simple calculations, and dynamic modeling.

The conceptual level can be accessed simply by reading the text. The numerical examples enable the second level: simple calculations and further conceptual insights. Competence in dynamic modeling can be enhanced by following the VisSim© examples available on the CD in the back of the book. The reader can open the examples in the VisSim Viewer available on the CD, which can also be installed on the hard disk via the Set-Up file included on the CD. A full working version of VisSim can be downloaded for trial from www.vissim.com/downloads/demos.html.

Similar examples in Simulink©, an excellent dynamic modeling software, are available for download at http://www.crcpress.com/e_products/downloads/default.asp

Preface

Thermodynamics, my professor J. Agrest used to say, should be called *thermostatics* instead. Indeed, most thermal system analysis assumes steady state or an indefinite time for processes to occur. The reality of energy systems is quite dynamic. Capturing, if only at an introductory level, the dynamic nature of energy supplies, conversion, and utilization is our objective.

Time and space are the conventional variables of engineering. Changing space coordinates requires energy. Energy use and time share one characteristic: irreversibility. Once energy is released, it flows, following gradients, until it becomes useless for human endeavor. Time seems to share with energy the characteristic of unidirectional flow. Tying energy systems and time by elucidating the time scales of energy systems has practical purposes: when applied to energy sources, the exercise may allow us to configure a number of virtual energy futures essential to the planning endeavor; when applied to technology, the exercise may allow us to evaluate the response time of prime movers, invariably a reason for their success; when applied to buildings, the exercise may allow us to identify energy saving opportunities.

Even before Plato's Academy, the Greeks were concerned with change, as evidenced by the passing of time. The energy outlook is changing fast as mankind searches for energy alternatives with minimal environmental consequences and acceptable cost. Like Aristotle, we search to identify a series of *telos*, namely, clear causes that can activate favorable change in our endeavor.

Studying the dynamics of thermal systems calls for suitable models derived from the laws of thermodynamics, and of rate relationships between flows and gradients. These laws are empirical and amount to a conceptualization of the natural world. They are reviewed in the first two chapters of this work. Many energy sources have been identified and developed, and we present their present state and potential to serve the needs of a growing world economy in Chapter 3, using dynamic analysis. To facilitate comprehension of the magnitude of different energy sources, we adopted tons of oil equivalent as the common unit.

Once released, energy is converted to useful forms. Hence, key fossil conversion technologies are described in Chapter 4. Dynamic models of selected prime movers show their potential to follow varying loads. All that is renewable technology must mesh with sources that are very much transient. Renewable conversion technologies and germane dynamic modeling aspects are covered in Chapter 5. Chapter 6 attempts to capture within the confines of a few pages, a seemingly inexhaustible topic: utilization technologies.

With paths from source to end use identified in previous chapters, we provide in Chapter 7 means to assess efficiencies from ground (or harvest) to end use. The results yielded by the method have many uncertainties, but it is our hope that the technique will help define dependable energy futures, or at least help discard some unfeasible ones.

Chapter 8 endeavors to capture the effects of energy use on the environment. Short- and long-term effects are discussed, with reference to global warming and radioactive decay.

Chapter 9 is a brief introduction to dynamic modeling. This topic deserves much more than a basic chapter, and many textbooks that offer a thorough treatment exist. The chapter is intended to assist readers who have no dynamic modeling experience. Those starting their own modeling effort will find the chapter helpful, but they must bear in mind that it cannot replace a good text on the topic.

Projections into the future have a way of eventually embarrassing their author. I hence wish I could have avoided Chapter 10, which is a description of energy technologies that might largely configure a sustainable energy future. Yet, a book of this nature, as my good friend Dr. R. L. Webb duly noted, would have been incomplete without those descriptions. Hence, the thoughts in Chapter 10 are intended to identify the *telos*, the hidden drivers that enable a desirable energy future for all.

MATLAB is a registered trademark of The MathWorks, Inc. For product information, please contact:

The MathWorks, Inc.
3Apple Hill drive
Natick, MA 01760-2098 USA
Tel: 508 647 7000
Fax: 508-647-7001
E-mail: info@mathworks.com
Web: www.mathworks.com

Acknowledgments

A complete list recognizing those who helped this book come about would exceed the possibilities of this space. I acknowledge, then, those of essential contributions first. For quiet, effective encouragement and editing, I thank T. L. Shock. For developing the Simulink© examples and preparing the electronic materials, I thank J. Perez-Blanco. For many of the ideas and insights that this book attempts to convey, I thank J. Agrest. For technical discussions and friendship, I thank D. Rose and R. Webb. For steadfast support in bringing this book to publication I thank J. Plant.

An energy book cannot be formulated in a vacuum, for it deals with a commodity that has become essential to the human endeavor. The many engineers I know who develop their activities around energy matters have also shaped this book. So have the students I taught, and so has my family. For countless ideas, discussions, insights, and observations, I thank and dedicate this book to all of them.

The Author

Dr. Perez-Blanco obtained his undergraduate degree at the University of Buenos Aires, and his M.S. and Ph.D. degrees at the University of Illinois, all in mechanical engineering. After 11 years at Oak Ridge National Laboratory dealing with thermal systems for energy conservation, he joined Penn State, where he has taught various aspects of mechanical engineering for 18 years, while conducting research in thermal systems, simultaneous heat and mass transfer, and gas turbine inlet cooling. His teaching has always converged into energy systems, particularly their dynamic modeling. He is a fellow of the American Society of Mechanical Engineers, teaches short courses on gas turbines for the International Gas Turbines Institute, and has developed an energy systems laboratory for his department.

1 The Laws

Because models, dynamic or static, must comply with the laws of thermodynamics, we review those laws in this chapter. A clear difference between mechanical and thermal energy is established here and related to fossil-fuel depletion. Temperature is presented as a gauge of thermal energy quality. Equations for dynamic modeling emanating from the laws considered in this chapter are derived in Chapter 2.

ENERGY

A system is a volume that we isolate (frequently only in our minds) for the purposes of analysis. Systems have energy, and energy is the capacity of the system to change itself, the environment, or both. For instance, if your system is a baseball, and it happens to have kinetic energy, it has the capacity to change its environment, say, break a window (and produce painful memories of the neighbor talking to your parents with ball in hand). If the system is the afflicted windowpane, the volume around it is penetrated by the ball, and upon shattering, the volume boundaries are crossed by glass shards and by the ball that ungratefully ends in the neighbor's hand (Figure 1.1).

Notice that the system boundaries can be in motion, as when the system is the ball, or can be stationary, as the volume that initially contains the windowpane. In the latter case, we call the system a control volume.

So, systems have boundaries, and, in the case of the ball at least, kinetic energy. And energy is the capacity to effect change, which, unlike in our example, does not necessarily have to produce an undesirable outcome. The state of a system is defined by the value of variables such as coordinates, temperature, and pressure. Energy has the capacity to change a system's state.

There are many agents with that capacity. In the case of the ball of mass m, the agent was kinetic energy. It is defined as KE, or

$$KE = \frac{m \cdot V^2}{2} \tag{1.1}$$

where V is the ball velocity.

Example 1.1 shows how the kinetic energy of the ball decreases as it breaks the window.

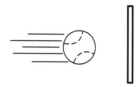

Before collision, the ball system
is in motion, and the window
pane system is stationary.

After collision, some of the ball's kinetic energy
is transferred to the glass. The windowpane
system is now empty.

FIGURE 1.1 Systems and boundaries.

EXAMPLE 1.1 ENERGY FOR WINDOW BREAKING

A ball strikes a glass windowpane as in Figure 1.1. The following data apply:

Ball mass	$m = 0.3$ kg
Initial velocity	$V_1 = 30$ m/s
Final velocity	$V_2 = 5.3$ m/s

Calculate

a. The energy consumed to break the window and sustain the process
   irreversibility.
b. The average rate at which the ball lost mechanical energy if the collision
   lasted an interval Δt.

$$\text{Collision duration} \qquad \Delta t = 0.013 \text{ s}$$

Solution

System: The ball.

Assumptions: No changes in the ball internal energy and no substantial heat transfer in the process of collision. The potential energy of the ball stays the same.
a. The change in kinetic energy is used to break the window. Hence

$$\Delta E = m \cdot \left(V_1^2 - V_2^2\right) \qquad \Delta E = 261.6 \text{ J}$$

is consumed to break the window.
b. The average rate of energy loss is simply

$$\frac{dW}{dt} \cong \frac{\Delta E}{\Delta t} \qquad \text{or} \qquad \frac{\Delta E}{\Delta t} = 20.1 \text{ kW}$$

The ball loses energy at a rate of 20.1 kW, or nearly twice the maximum rate at which
the normal household can receive energy from the grid.

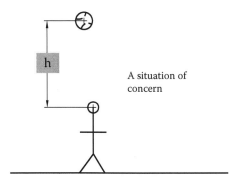

FIGURE 1.2 A ball above your head has potential energy.

From experience, we know that a ball way above our heads also has the capacity to do work, and we label that capacity the potential energy or *PE*:

$$PE = m \cdot g \cdot h \tag{1.2}$$

where h is the elevation the ball has over the head, and g is the gravitational constant.

Now, if we animate the drawing of Figure 1.2 in our minds, we find that, as the ball falls, it loses elevation but picks up speed. This is a particularity of energy: one can think of it as being convertible from one form into another—in the example, from potential into kinetic energy.

Yet another form of energy is found necessary for the proper study of thermodynamics—the internal energy. As with many concepts, it is easier to establish what internal energy is not rather than what it is. Internal energy is the kinetic and potential energy of the molecules that make up the mass inside the system; it is not the kinetic or potential energy of the whole system but that of the molecules that make up the system. Imagine two baseballs connected with a spring, hurled so that they travel together while also oscillating. Our system is now the two balls. The center of mass of the system follows the parabolic trajectory that a single ball with the mass of the two balls would follow, but the kinetic energy of the system is the sum of the energy of oscillation plus that of translation. If the balls, in addition to vibrating, were to rotate around the center of mass (such rotational motion is not depicted in Figure 1.3), then they would store more energy in the form of

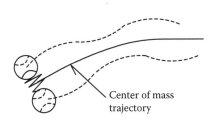

FIGURE 1.3 Center of mass trajectory and individual trajectories.

rotational kinetic energy. The translational kinetic energy of the system is in a way different from the vibrational/rotational energy of the balls around the center of mass. In this coarse analogy, the former would be the kinetic energy of the system, and the latter the internal or stored energy because it is contained within the system.

We need to distinguish via symbols between the three energies discussed so far. In addition, we define E, the total energy of the system, equal to the sum of the other three:

$$E = U + KE + PE \qquad (1.3)$$

  E: system energy
  U: internal energy
 KE: kinetic energy
 PE: potential energy

So far, we only have energy in the form of motion (kinetic) or position (potential), animating the whole system or internal parts of it. The system, however, has boundaries. For the energy of a system to increase or to decrease, energy must be able to cross the boundaries. Example 1.2 shows how kinetic energy can be related to internal energy.

EXAMPLE 1.2 INTERNAL ENERGY

Two balls are joined by a spring, as shown in Figure 1.3. An oscillatory motion is imparted to the pair, which is then hurled away. As it is thrown, a rotational motion is communicated to the assembly.

$$\text{Ball mass} \quad m = 0.3 \text{ kg}$$

$$\text{Spring mass} \quad ms = 0.06 \text{ kg}$$

At the instant of interest, the assembly has

$$\text{Vibrational energy (translation plus elastic energy)} \quad Ev = 2.3 \text{ J}$$

$$\text{Rotational energy} \quad Er = 3.4 \text{ J}$$

$$\text{Velocity of center of mass} \quad Vcm = 12.2 \text{ m/s}$$

Calculate

  a. The kinetic energy due to translational motion of both balls
  b. The energy internal to the system formed by the two balls

Solution

System: Both balls and spring.
Assumptions: We make the analogy that the kinetic energy of the center of mass is the kinetic energy of the system, whereas the rotational and vibrational energies represent the internal energy.

a. For the chosen system, we take the kinetic energy of the total mass

$$KE = \frac{1}{2} \cdot (2 \cdot m + ms) \cdot Vcm^2 \qquad KE = 49.1 \text{ J}$$

b. The internal energy to this system would be the sum of the vibrational and rotational energies

$$U = Er + Ev \qquad U = 5.7 \text{ J}$$

ENERGY IN MOTION

One way energy crosses boundaries occurs when mass crosses the boundaries, as in the example of the broken window. Each mass particle carries energy with it. However, energy crossing the boundaries of a system does not necessarily need to be transported by mass. The concept that energy transport can occur via mechanisms other than mass crossing the boundaries is somewhat of a newcomer to traditional physics. Not more than 150 years ago, people still needed to invoke an imaginary fluid (called "caloric") to explain some energy flows. Yet, the energy transport occurs not through the flow of some mysterious fluid but as a consequence of differences in two key variables: temperature and pressure. Let us consider the former.

Temperature essentially measures how hot a body is with reference to a convenient zero. What the reference is matters little to our discussion, although there is one absolute zero, a topic of great insight and beauty. What matters (Figure 1.4) is that if one puts a cold brick on top of a hot brick and wraps the assembly with an insulating blanket, eventually the hot brick becomes colder and the colder brick becomes hotter.

Eventually is the key word here. With enough time, the system composed of the two bricks reaches the same uniform temperature, and once that condition is reached, it will stay that way for an eternity. Hence, the dynamics of the system define the time during which changes will take place. Why does the temperature of each block change? Our intuition (and measurements) would tell us that the mass of each brick does not change.

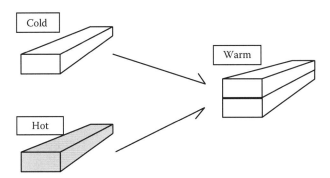

FIGURE 1.4 Bricks at initially different temperatures eventually come to equilibrium.

Yet, the temperature of each brick changes, and if we choose to regard each brick as a system, clearly the cold one gains energy and the hotter one loses energy. Wait a minute, you say, how could the bricks gain or lose energy if they are not moving?

The bricks are not moving (setting aside any expansion or contraction that they may be experiencing), but the molecules of the hot one lose some of the vibrational or rotational kinetic energy animating them, whereas the molecules of the cold one gain in the same type of energy. Hence, temperature is a good way to measure the internal energy of each brick. One wonders now how the energetic molecules transfer their energy to the less energetic ones. The answer is through energy flow across the boundaries of each brick, and the interfacial process is well beyond our scope. The flow ceases when the bricks reach thermal equilibrium.

The energy in transit due to a temperature difference is called heat. The symbol Q denotes the heat transferred across the system boundaries during the time interval of interest. If we define our system to be the hot brick, then, during the time it takes to go from temperature T_1 (hot) to temperature T_2 (cold), we know two things to be true:

a. The change in brick internal energy is equal to the heat lost:

$$\Delta E_{12} = Q_{12} \tag{1.4}$$

b. The heat lost as the brick evolves from 1 to 2 depends on the mass m, the specific heat c, and the temperature change as

$$Q_{12} = m \cdot c \cdot (T_2 - T_1) \tag{1.5}$$

The product $m \cdot c$ is often called C, the heat capacity of the brick. Equation 1.5 simply says that the greater the mass or the specific heat or the temperature difference, the greater the heat loss will be. Because $T_2 - T_1$ is negative, the energy change is also negative (Equation 1.4), and we say that the system loses energy. Similar equations for the initially cold brick would show that its internal energy increases although the kinetic and potential energy of both bricks remain unchanged.

Example 1.3 shows how heat transferred can be related to internal energy.

EXAMPLE 1.3 HEAT AND INTERNAL ENERGY

A metal ingot (Figure Example 1.3.1) increases its temperature from 285 K to 300 K, receiving heat. The process takes place during a time interval Δt.

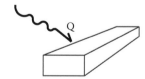

FIGURE EXAMPLE 1.3.1 Metal ingot.

a. Calculate the change in internal energy of the ingot for the following properties.
b. Calculate the average heat flow crossing the ingot boundary during the time interval given.

$$T_1 = 285 \text{ K} \qquad T_2 = 300 \text{ K}$$

$$\text{Mass} = 0.5 \text{ kg} \qquad c = 1800 \frac{\text{J}}{\text{kg} \cdot \text{K}} \qquad \Delta t = 15 \text{ min}$$

Solution

System: Ingot.
Assumptions: The metal ingot temperature is uniform, and heat only crosses the system boundaries.

a. The change in internal energy equals the heat that went into the ingot. The heat over the interval given is obtained from Equation 1.5 as

$$Q = mass \cdot c \cdot (T_2 - T_1) \qquad Q = 13500 \text{ J}$$

Hence, the change in internal energy is

$$\Delta E = Q = 13500 \text{ J}$$

b. The average heat flow is calculated as the quotient of the heat exchanged and the time for exchange:

$$\dot{Q} = \frac{Q}{\Delta t} \qquad \dot{Q} = 15 \text{ W}$$

How fast will a system exchange heat across its boundaries? This is a matter of heat transfer, a rather empirical science. At this point, suffice it to invoke the cooling law (also from Newton) for convection, which indicates that the heat loss between a solid surface (which coincides with the boundaries of the system) and a surrounding fluid is proportional to the temperature difference between the system boundaries and the surroundings, multiplied by a heat transfer coefficient hc and by the surface area:

$$\frac{dQ_{lost}}{dt} = hc \cdot A_{sys} \cdot (T_{sys} - T_{atm})$$

Empirically, this makes sense: the larger the area, the greater the heat loss rate. Also, the larger the heat transfer coefficient (a function of the fluid velocity), the faster heat will flow. However, the crucial dependence is on temperature difference: temperature determines the direction of heat flow.

Example 1.4 shows how heat transfer rates vary with time.

EXAMPLE 1.4 COOLING RATE

A hot ingot is cooling down. Heat loss occurs by convection, that is, the rate of heat loss is proportional to the temperature difference between the ingot and the environment, assumed to remain at constant temperature. The heat capacity, surface area and initial temperature of the ingot, the environment temperature, and the average (constant) heat transfer coefficient are given:

Ingot data

Heat capacity	$Ci = 3.10^6$ J/K
Initial temperature	$Ti_1 = 420$ K
Surface area	$Ai = 0.4$ m²

Other information

Environment temperature	$T_{atm} = 380$ K
Heat transfer coefficient	$hi = 106 \dfrac{W}{m^2 \cdot K}$

Calculate

  a. The initial rate of heat loss from the ingot.
  b. The equation for the temperature variation of the ingot versus time.
  c. Plot the ingot temperature and heat loss versus time. Comment on the shape of the traces as related to the rate of heat loss.

Solution

System: The ingot.
Assumptions: The predominant mode of heat transfer is convection, as indicated in the problem statement

  a. The initial rate of thermal energy loss is the heat transfer coefficient × the ingot area × the initial temperature difference:

$$\frac{dQ}{dt} = hi \cdot Ai \cdot (Ti_1 - T_{atm})$$

from which we obtain the initial rate of heat transfer as

$$hi \cdot Ai \cdot (Ti_1 - T_{atm}) = \frac{dQ}{dt} = 1700 \text{ W}$$

  b. In addition to some heat transfer memories, the reader must invoke here some of his or her calculus:
    The heat lost by the ingot at any point in time is given by

$$\frac{dQ}{dt} = hi \cdot Ai \cdot (Ti_1 - T_{atm}) = -Ci \cdot \frac{dTi}{dt} \qquad \text{(a)}$$

The negative sign in front of Ci indicates that the temperature of the ingot decreases as long as its temperature is above that of the environment. Manipulation of Eq. (a) and integration between time 0 ($T_i = T_{i1}$) and the final time tf ($T_i = T_{i2}$) yields:

$$\int_{Ti}^{Ti_2} \frac{1}{(Ti - T_{atm})} \cdot dTi = \int_{0}^{tf} -\frac{hi \cdot Ai}{Ci} \cdot dt$$

Performing the integration gives

$$\ln\left(\frac{Ti_2 - T_{atm}}{Ti_1 - T_{atm}}\right) = -\frac{hi \cdot Ai}{Ci}tf$$

Rearranging to express Ti_2 as a function of time, we obtain

$$Ti_2(tf) = (Ti_1 - T_{atm}) \cdot \exp\left[-\frac{hi \cdot Ai}{Ci} \cdot (tf)\right] + T_{atm}$$

The rate of heat loss ($HLR\ [tf]$) is simply

$$\frac{dQ}{dt} = HLR(tf) = (Ti_1 - T_{atm}) \cdot \exp\left[-\frac{hi \cdot Ai}{Ci} \cdot (tf)\right] + T_{atm}$$

c. We plot the ingot temperature to obtain an exponential decrease with time (Figure Example 1.4.1).

The heat loss rate ($HLR[tf]$) is high initially, but decays with time (Figure Example 1.4.2).

Observations:
   a. The ingot is not very hot, but it is massive; it takes a long time (about 100 h) for it to cool to room temperature.

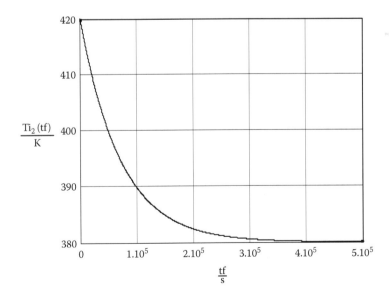

FIGURE EXAMPLE 1.4.1 The ingot temperature decreases quickly initially and approaches its final value asymptotically.

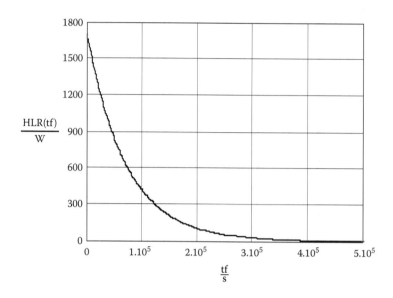

FIGURE EXAMPLE 1.4.2 The ingot heat loss decreases rapidly as the temperature approaches its final value, which takes a long time.

b. The ingot temperature decreases exponentially with time, and hence its rate of decrease diminishes with time.
c. The heat loss is proportional to the ingot–atmosphere temperature difference, and hence the heat loss rate decreases as time progresses.
d. There is no heat transfer when there is no temperature difference.

Heat transfer changes the temperature but not the state of motion of solids. For gases and liquids, as well as solids, internal energy changes can come about because of heat transfer. There are cases in which temperature changes originate buoyancy forces that result in fluid motion, but more often than not, heat transfer merely changes the temperature and, with it, the internal energy of the system. The state of motion of mass is ordinarily changed by another form of energy in transfer due to pressure differences.

To see how pressure differences enter the picture of energy in motion, consider the definition of mechanical work. An infinitesimal amount of work dW is done when a force $\vec{F}$ moves along a path $d\vec{s}$, with a component of that force coinciding with the direction of the path. Using the dot product for the two vectors, we have

$$dW = \vec{F} \cdot d\vec{s} \tag{1.6}$$

Starting with Equation 1.6, we now picture a pressure p acting on the finite area ΔA on the system boundary (Figure 1.5). Since the pressure is normal to the surface on which

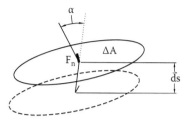

FIGURE 1.5 Differential of mechanical work.

it acts, the net force acting on ΔA is also normal to the surface, and it is given by

$$F_n = \Delta A \cdot p$$

The area ΔA is displaced to the new position indicated by the dotted line in a certain time interval. According to Equation 1.6, the work done by F_n is

$$dW = \vec{F} \cdot d\vec{s} = F_n \cdot ds \cdot \cos(\alpha) = p \cdot \Delta A \cdot ds \cdot \cos(\alpha)$$

Since the product $\Delta A \cdot ds \cdot \cos(\alpha)$ is just the volume $dVol$ swept by ΔA during the displacement, then work done is

$$dW = p \cdot dVol \tag{1.7}$$

and the power developed is simply

$$\frac{dW}{dt} = p \cdot \frac{dV}{dt} \tag{1.7a}$$

Now it is easy to imagine what a certain amount of work does to the system: if the boundary is rigid, then the work done will increase the kinetic energy of the system. For instance, if a force is applied to our ball (and it is rigid and friction is negligible), the ball will accelerate until the work done equals the kinetic energy. If the system is compressible (i.e., a gas inside a cylinder fitted with a piston, the proverbial thermodynamic example), then work done increases the temperature and the pressure of the gas. This is what happens when one uses a manual pump to inflate a bicycle tire (Figure 1.6): the piston performs work on the gas, raising its pressure (so that it can flow into the tire) and its temperature (which is unfortunate because one would want all the work to go into raising the pressure, not the temperature of the gas).

What if the external pressure exceeds the pressure of the gas inside the piston? The work done on the system is still given by the integral of Equation 1.7 between the initial and final states, but the work done by the external pressure will exceed the

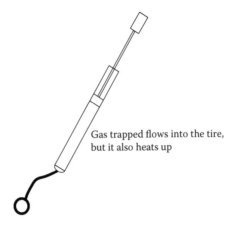

Gas trapped flows into the tire, but it also heats up

FIGURE 1.6 Pump and work done on the air.

work done by the piston on the gas. Calculating the integral

$$W = \int_1^2 dW = \int_1^2 p \cdot dVol \qquad (1.8)$$

for the work done proves challenging since in most cases the pressure distribution inside the gas is unknown, especially if the process occurs very rapidly. Insightful approximations to this long-standing problem are common.

Example 1.5 shows how expansion work and power depend on the choice of system.

EXAMPLE 1.5 EXPANSION WORK

A child is inflating a rubber balloon. The radius changes from $r1$ to $r2$ with the radius as function of time given as $r(t)$. The pressure p_i inside the balloon, the force per unit length S exerted by the rubber, and the external pressure p_{atm} are related by

$$p_i = p_{atm} + \frac{S}{r}$$

The following is known:

$$p_i = p_{atm} + \frac{S}{r}$$

$$S = 100 \frac{N}{m}$$

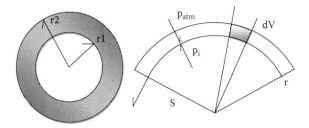

FIGURE EXAMPLE 1.5.1 Balloon and tension due to stretching.

As the balloon is inflated from $r1$ to $r2$, the value of S can be assumed constant, and the balloon can be assumed to be spherical (Figure Example 1.5.1)

$$r1 = 0.15 \text{ m} \qquad r2 = 0.19 \text{ m}$$

The radius varies according to the function of time:

$$r(t) = r1 + 0.01 \cdot \frac{m}{s} \cdot t$$

Calculate

a. The time taken to go from $r1$ to $r2$.
b. The rate of work done by the balloon on the atmosphere as function of time. Plot this rate.
c. The rate of work done by the internal air on the balloon and the atmosphere. Plot both the power in b and in c and comment.

Solution

System: (a) The air inside plus the balloon. (b) The air inside the balloon only.
Assumptions: S is constant, and the balloon is spherical.

a. The time required to get from $r1$ to $r2$ is simply 4 s, for

$$r2 = r1 + 0.01 \cdot \frac{m}{s} \cdot t2$$

$$t2 = \frac{r2 - r1}{0.01 \cdot m/s}$$

$$t2 = 4 \text{ s}$$

b. We use the expression for work (Equation 1.7a) with known pressure:

$$\frac{dW}{dt} = P_{atm} \cdot \frac{dVol}{dt}$$

The rate of volume change is

$$Vol = \frac{4}{3} \cdot \pi \cdot r^3$$

$$\frac{dVol}{dt} = 4 \cdot \pi \cdot r^2 \cdot \frac{dr}{dt}$$

and the rate of radius change is, taking the time derivative of $r(t)$,

$$\frac{dr(t)}{dt} = 0.01 \, \frac{m}{s}$$

The power developed by the balloon then is, using the time derivatives of Vol and of r,

$$\frac{dW}{dt} = p_{atm} \cdot \frac{dVol}{dt} = p_{atm} \cdot 4 \cdot r(t)^2 \cdot 0.01 \, \frac{m}{s} = Power(t)$$

This equation is plotted in Figure Example 1.5.2.

c. In this case, for the air inside the balloon, we have

$$\frac{dW}{dt} = \left(p_{atm} + \frac{S}{r} \right) \cdot \frac{dVol}{dt} = \left(p_{atm} + \frac{S}{r} \right) \cdot 4 \cdot r(t)^2 \cdot 0.01 \, \frac{m}{s} = Powerb(t)$$

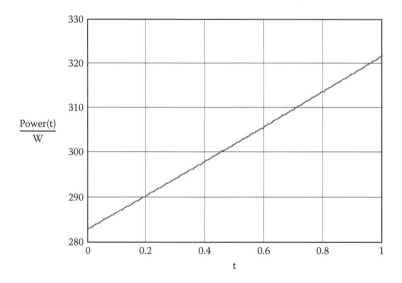

FIGURE EXAMPLE 1.5.2 The inflation power increases as the volume increases.

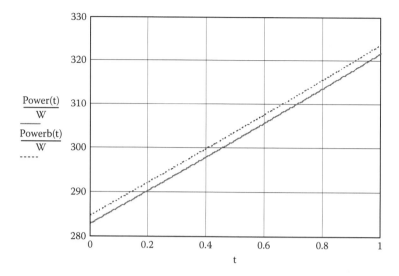

FIGURE EXAMPLE 1.5.3 Incorporating the tension of the balloon material increases the calculated inflation power.

The air inside the balloon must do work on the atmosphere and to expand the rubber of the balloon. Hence, the work done by the system air inside is greater than the work done on the atmosphere, Figure Example 1.5.3.

THE SECOND LAW

If you put excess toothpaste on the toothbrush, the thought of returning the excess paste to the tube does not even cross your mind. It is impossible. Of course, you could visit Mastodon's and Bros. to get a small motor and couple it to a pump capable of handling viscous fluids to force the paste into the tube. This would require an inordinate amount of work: from that required to get to Mastodon's to manufacture the pump to the electric power that the motor would consume. The Second Law relies on the observation that all natural processes occur preferentially in one direction; reversing matters to their original condition requires work.

More importantly, when a process occurs spontaneously as a consequence of an existing lack of equilibrium, some work is lost. The best that one can hope for is to capture some of that work, but (alas!) not all of it. Consider for instance a mass of water in a lake on top of a mountain, a problem that has more engineering flavor than the toothpaste scenario. The water is not in equilibrium with the water down in the valley (Figure 1.7). It can flow via the river to the valley, or you can install a turbine and get some work out of this disequilibrium. However, because the water runs into rocks, bends, logs, and proverbial fish swimming upstream, it never picks up as much speed as it could, and your turbine never gives you as much work as you expected. Of course, you could install a pipeline from the lake at the top to the lake in the valley and accelerate the water as much as possible before hitting the turbine. This is a

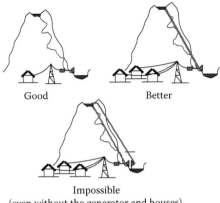

Good   Better

Impossible
(even without the generator and houses).

FIGURE 1.7 Possibilities for harnessing the disequilibrium between water in the mountain-top lake and water in the valley.

much improved situation, but friction in the pipeline will always retain some small fraction of the possible work.

How are we so sure that some work will be lost to friction in the pipe? Because, to move flow inside a pipe requires a pressure drop, as shown in Figure 1.8. Water is lazily at rest inside a pipe. The pressure is the same at points 1 and 2. You open the tap, and 2 is communicated to the lower atmospheric pressure. Water flows inside the pipe and flows down the drain. Water flows because (after the tap is opened) the pressure is larger at point 1 than at point 2. The higher pressure at 1 is necessary to overcome the frictional resistance to motion. Reversing your actions would require pumping water (the same water, mind you) to the original pressure inside the pipe, with the corresponding work expenditure to run the pump.

Friction, then, will rob us of work that the system could have produced. This is the way things are, and this is the way they have been for a long time (Figure 1.7). Because of friction, we cannot use the power of the turbine to run a pump to restore the water to its original position so that it can run the turbine again. We cannot even break even: friction destroys work, and the pump will never be capable of restoring the water to the original elevation. Bummer!

Friction is a form of irreversibility. Irreversible processes are those that defy any attempt at restoring matters to their original condition. Irreversibility can take many

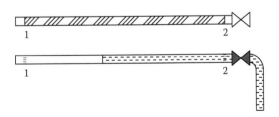

FIGURE 1.8 Water flows when the valve is opened.

forms, but it seems to be the trademark of all natural processes. Work is always required to reverse a process, but the production of that work entails other irreversible processes. Only in your wildest dreams are engineering systems reversible. No one has come up with any, but this is not for lack of trying. What this means is:

One will never get as much work from a real process as from a reversible one (which is a construct of our minds anyway).
Restoring a system and environs to their original condition after a process has occurred is impossible.

Thus stated, the Second Law can be read as the creed of a pessimistic and negative bunch, badly in need of antidepressants which will do them no good anyway. This is not the case. The Second Law simply states matters as they stand, and no amount of wrangling will change its solid reality: in this world, work is lost to irreversibility.

Example 1.6 discusses further the consequences of irreversibilities and power produced.

EXAMPLE 1.6 THE SECOND LAW: EXAMPLES AND REDUCED POWER DUE TO IRREVERSIBILITY

The Second Law maintains that reversible processes do not exist in the physical world, irreversible processes being the norm. Consequently, examples of the former are difficult, if not impossible, to find. Yet, over the human time scale, reversibility exists in some domains. The system formed by atoms and their electrons appears, for all practical purposes, reversible if not easily harnessed. However, the atomic processes that are relatively easy to harness imply irreversible decay; fission is irreversible.

Planetary motion, another unharnessed energy form, seems to decay so slowly that it could be construed as to be nearly reversible. This mechanical motion seems impervious to the passing of time in the human scale. Still, light and matter do not flow back out of black holes in the Universe, which might indicate that there is a definite direction to time even in a space so seemingly unbounded that our minds have trouble appraising it. Similarly, the order of day and night on Earth has never been reversed such that sunlight reversed its direction and flowed back to the Sun to split He atoms into Hydrogen ones. Irreversibility is all around us.

Here we give a few parameters for a water turbine system (Figure Example 1.6.1), showing that, in real processes, less work is produced due to irreversibilities.

Water flows from a lake into a penstock and through a turbine. For the given data, for reversible and actual (i.e., irreversible processes), identify the cause of irreversibility. Comment on each step.

1. Possible real case $p1 = 100000$ Pa $V1 = 3$ m/s $h1 = 100$ m
                              $p2 = 88532$ Pa $V2 = 3$ m/s $h2 = 2$ m
                              $p3 = 100000$ Pa $V3 = 2$ m/s $h3 = 0$ m
                              $Power = 7.48$ kW
2. Reversible case $p2r = 100000$ Pa $V2r = 3$ m/s $h2r = 2$ m
                              $p3r = 100000$ Pa $V3r = 3$ m/s $h3r = 0$ m
                              $Power = 8.65$ kW

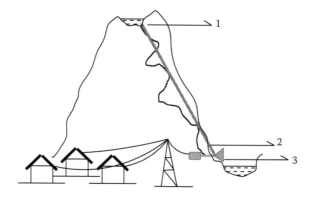

FIGURE EXAMPLE 1.6.1 A small hydro plant.

Solution

From pt 1 to pt 2:

System: Pipe from 1 to 2, water flowing in and out at a steady rate.

The difference between the reversible and irreversible case is one of pressure, the velocities and elevations being the same: $p2 < p2r$ because the real fluid requires a pressure drop to move inside the pipe. Therefore, the pressure at the turbine inlet is smaller than it could have been for ideal flow, and even an ideal turbine would produce less power than it could have.

From pt 2 to pt 3:

System: Turbine from 2 to 3, water flowing in and out at a steady rate, power delivered at steady rate.

The ideal turbine produces more power than the real one because

   a. It receives a higher head pressure.
   b. It extracts all the kinetic energy from the flow ($V3r = 0$ m/s).
   c. It has no friction losses in the internal parts and in the fluid flow.

The real turbine is not as well designed, but even if it were redesigned to capture all the flow kinetic energy ($V3~0$ m/s, very large machine) and to work with a reversible pipeline from 1 to 2, the turbine would still have losses due to friction, and its power output (for a mass flow rate of 10 kg/s) would be 8.2 kW, still below the reversible one.

What is the typical fate of lost work? It becomes heat that flows to the lowest temperature available, typically the environment. A clear example is the braking system of some transportation systems. To slow down a car speeding downhill, the brakes could be applied, reducing the speed by friction between the rotors and braking pads. All that good kinetic energy is converted into internal energy of the rotors and pads, and flows out of them as heat. Brakes are known to overheat if used too frequently because all they do is to take kinetic energy (it is a form of mechanical energy) and convert it into heat via friction. This can be avoided (only partially, alas!) with recuperative braking systems.

Among recuperative systems, the author's favorite ones are the San Francisco cable cars, although recuperative systems are applied to automobiles and bicycles as well. When a cable car goes downhill, the driver can apply the brakes or he can firmly grasp the cable that runs under the car. Then, the car will run at the cable speed, and the cable will absorb its decrease in potential energy. That decrease is used by another car hooked to the same cable, but going uphill, to raise its potential energy. In a more modern example, many electric train systems have recuperative braking, sending power to the net as they slow down. With enough trains on the net, when one slows down, another should be speeding up and using the recovered energy, plus some input from the power plant. Recuperative systems are not immune to the curse of irreversibility, and they will require external power input for continued operation.

THE FIRST LAW

So far, we have defined the energy of a system as basically stored in one way or another in the system, and heat and work as the means of transferring energy across the system's boundaries. Graphically, we show that energy is made up of kinetic, potential, and internal energy, whereas work and heat respond to different gradients and are energy in motion. Hence, they are shown as different sets in Figure 1.9.

If for the moment heat as a form of energy transfer is ignored, we have only energy and work as players. For instance, when a cart (let us call it the Thermal cart) is pushed to communicate kinetic energy to it, it follows that the gain in kinetic energy is equal to the work done. The differential of work done is (Figure 1.10), at any instant,

$$dW = -F \cdot ds \qquad (1.9)$$

where the minus sign indicates that work received by the system is considered negative. The rate at which work is done is now the power input to the Thermal cart:

$$P = \frac{dW}{dt} = -F \cdot \frac{ds}{dt} = -F \cdot V \qquad (1.10)$$

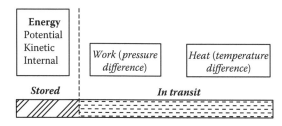

FIGURE 1.9 The three forms of energy.

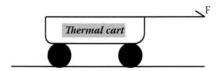

FIGURE 1.10 The thermal cart is accelerated by a constant force F.

As work is done on the Thermal cart, its velocity increases. A differential of work equals the increment in kinetic energy. Since the kinetic energy is

$$KE = \frac{1}{2} \cdot m \cdot V^2$$

its differential is

$$dKE = m \cdot V \cdot dV$$

and the rate at which KE increases is simply

$$\frac{dKE}{dt} = m \cdot V \cdot \frac{dV}{dt}$$

Newton's Second Law indicates that the product $mass \cdot dV/dt$ is equal to F, the applied force, and hence

$$\frac{dKE}{dt} = F \cdot V$$

Or, by virtue of Equation 1.10, we get

$$\frac{dKE}{dt} = -\frac{dW}{dt} \qquad\qquad (1.11)$$

From Equation 1.11, the rate of work done equals the rate of change of kinetic energy, but one is positive while the other is negative. Hence

$$\frac{dKE}{dt} + \frac{dW}{dt} = 0$$

It would be great if we could generalize and stop here; however, there is more. Say that you help the Thermal cart climb a hill, raising its kinetic and potential energy. Then, at any instant, the rate of work done equals the rate of gain (loss) of potential and kinetic energies, and hence

$$\frac{dKE}{dt} + \frac{dPE}{dt} + \frac{dW}{dt} = 0$$

If we were now done, the Second Law would be but a figment of our imagination. We just labored through a couple of pages to assert that irreversibility is an unavoidable fact of technical life. Irreversibility requires consideration of heat transfer, which we have so far excluded from the discussion. We include it now. Hence, imagine that the friction of the wheels on the ground and of the bearings heats the Thermal cart. We do not know how much of the work we do is against friction; we only know that it will inevitably rear its ugly head and make us work harder, unless our goal in life is to heat the atmosphere. The heating is local, but let us simplify and agree that some average temperature would define the internal energy of the Thermal cart. At any instant then, in addition to variations of KE and PE, our internal energy U will also vary at a rate given by

$$\frac{dU}{dt} + \frac{dKE}{dt} + \frac{dPE}{dt} + \frac{dW}{dt} = 0$$

We are almost done. We now include a crucial consideration. Our system is the Thermal cart, with boundaries that coincide with it, somewhat like the plastic wrapping that covered the cart when it was new. As parts of the Thermal cart heat up, they become hotter than the surrounding air, and some heat (dQ/dt) will flow into the air. The heat loss dQ/dt we take as negative. Hence, we now have

$$\frac{dU}{dt} + \frac{dKE}{dt} + \frac{dPE}{dt} - \frac{dQ}{dt} + \frac{dW}{dt} = 0 \qquad (1.12)$$

From this equation, it is clear that if the cart has bad bearings and soft wheels, and your intent is to communicate potential and kinetic energy at a certain rate, you will have to work harder to meet the heat loss and the increase in internal energy that come about because of irreversibility. Aware of all these implications, and keeping in mind that $E = U + KE + PE$, we can now write, using Equations 1.3 and 1.12,

$$\frac{dE}{dt} = \frac{dQ}{dt} - \frac{dW}{dt} \qquad (1.13)$$

Of course, as you keep on racing the Thermal cart, its occupant may need to effect repairs, say in the nature of welding a rapidly failing part. Certainly, the torch heat will heat the cart, and Equation 1.13 will reflect this with a heat gain, not a heat loss. But simply applying a torch to the cart will not make your job of pushing it any easier; the heat input will increase the cart's internal energy but not its kinetic or potential energy.

Equation 1.13 indicates that energy is conserved, and that internal energy, heat, and work are interchangeable. The First Law fails to indicate that, in all actual systems, some of the work input will end up as heat and not as energy, and conversely that some of the work output will be dissipated as heat before it can be used.

Example 1.7 discusses sustainability of a simple conversion process and the First Law.

EXAMPLE 1.7 FIRST LAW, USING INTERNAL ENERGY

A closed system is operating at a certain instant delivering work at a rate dW/dt, inputting heat at a rate dQ/dt, and reducing its internal energy at a rate dE/dt.

a. Calculate dE/dt for the following values.

$$\frac{dW}{dt} = 32 \text{ kW} \qquad \frac{dQ}{dt} = 17.3 \text{ kW}$$

b. What would be the consequence of attempting to perform this operation for a long time regarding the system temperature? (The system does not have kinetic or potential energy, only internal energy.)

Solution

System: The given closed system

a. The First Law reads:

$$\frac{dE}{dt} = \frac{dQ}{dt} - \frac{dW}{dt} = 17.3 - 32 = -14.7 \text{ kW}$$

b. For a system with only internal energy, the main consequence of extended operation in this mode would be that the energy stored (and the temperature) in the system would decrease, eventually becoming null, and the process could not be sustained any longer.

Example 1.8 discusses power and heat rates in an engine converting thermal energy into power.

EXAMPLE 1.8 FIRST LAW OVER A CYCLE

An engine works in a cycle with a period Ti. At 2000 rpm and steady state, the power output as a function of time is given as

Period $Ti = 0.03$ s

$$\frac{dW}{dt} = Wdot(t) = \left[33 + 150 \cdot \sin\left(\frac{2 \cdot \pi}{Ti} \cdot t\right) + 35 \cdot \sin\left(\frac{4 \cdot \pi}{Ti} \cdot t\right) + 21 \cdot \sin\left(\frac{5 \cdot \pi}{Ti} \cdot t\right) \right] W$$

The heat rate exchanged by the engine as a function of time is

$$\frac{dQ}{dt} = Qdot(t) = \left[35.674 + 205 \cdot \cos\left(\frac{2 \cdot \pi}{Ti} \cdot t\right) \right] W$$

For this cycle,

  a. Plot the power output, the heat exchanged, and the rate of change of energy of the engine. Comment on the plots and the direction of the energy exchanged.
  b. Calculate net heat and net work over a cycle.
  c. Give the net energy change of the engine over a cycle.

  a. The rate of energy change is, by the First Law,

$$\frac{dE}{dt} = Edot = \frac{dQ}{dt} - \frac{dW}{dt} = Qdot(t) - Wdot(t)$$

Plotting the functions just defined, we obtain the graph Figure Example 1.8.1.

The power developed (solid trace) is positive in part of the cycle, whereas the engine receives power from an external source during the rest of the cycle. Only if the integral of the power over a cycle is positive will the engine deliver useful power. The engine receives heat (*Qdot* is positive) during a large part of the cycle. However, there is heat rejection when *Qdot* is negative.

Since in a cycle (by definition) the engine returns to its original condition, then, over a cycle, it must be true that

$$\Delta E = Q - W = \int_0^{Ti} Qdot(t)\,dt - \int_0^{Ti} Wdot(t)\,dt = 0$$

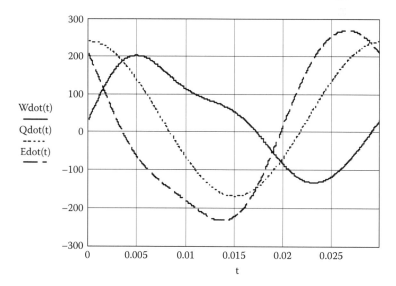

FIGURE EXAMPLE 1.8.1 Traces of power, heat rate, and energy for one period.

which implies that the net heat input is converted into the net work output.

    b., c. Integrating the given functions for *Qdot* and *Wdot*, we obtain

$$Qnet = \int_{0}^{0.03} Qdot(t)\,dt = 1.07 \text{ W}$$

$$Wnet = \int_{0}^{0.03} Wdot(t)\,dt = 1.07 \text{ W}$$

The net energy variation over a cycle should be null:

$$\Delta E = Qnet - Wnet \cong 0 \text{ W}$$

IF ENERGY IS CONSERVED, WHY WORRY ABOUT ENERGY SUPPLIES?

The First Law tells us that the sum of energy rates (Equation 1.13) is constant, implying that energy is transformed, not destroyed. The Second Law tells us that there is a preferential, spontaneous direction of transformation, from mechanical, potential, or kinetic energy to heat. Further, even heat is transformed, as it flows spontaneously from hot to cold, that is, the temperature at which it is available becomes lower and lower as it flows.

Energy supplies are susceptible to depletion because all energy tends to end up as low-temperature energy. Once energy is at low temperature, it is impossible to recycle it into high temperature again without additional energy expenditure.

It gets worse yet, for most of the low-temperature heat leaves the planet, further complicating any recycling scheme. Unlike, for example, metals (the mass of which is constant on Earth, the occasional space probe or meteorite notwithstanding), energy is continually received from space and released on Earth from combustion and nuclear processes. The amount rejected to space roughly equals the sum of the energy received and released by humans, but the rejection occurs at a lower temperature. Hence, energy constantly undergoes the irreversible process of flowing to a colder temperature away from the Earth.

The incoming energy does not generate fossil fuels fast enough to satisfy our appetite for them, nor can it regenerate nuclear fuel. Hence, there is some reason to plan (worry seldom accomplishes much) for the eventual exhaustion of resources. Additionally, planning may avoid some of the unwanted consequences of fossil and nuclear fuel by-products, which is another aspect of irreversibility.

TEMPERATURE: CAN IT GET ANY COLDER?

Absolute temperature as a search topic currently yields $1.7 \cdot 10^6$ search results on the Web—too many for a human to check out. At 3 min per entry, 10 years of human life would go researching this topic that many find fascinating. Of absolute temperature, the dictionary [1] says little, although it contains a good definition of temperature in lay terms: "Physical magnitude that objectively characterizes the subjective sensation of hot or cold arising from the contact with a body." This anthropomorphic definition is apt, in that it is a human that feels what is cold or hot, albeit that a thermometer yields the objective measurement. In any case, a thermometer requires a reference: there is a zero in every scale, but it is usually arbitrary. (The freezing point of water is the zero in the Celsius scale, whereas the temperature of a mixture of sea salt, ice, and water is the zero in the F scale.)

Many humans may have wondered how much colder it can get than say 0°C (zero Centigrade, 32°F). When physicists took the matter up, they found that indeed there is an absolute zero for temperature. A brief explanation follows. We have indicated that temperature is a good gauge of internal energy (for solids and liquids at least, and sometimes for gases as well) and that internal energy is the "catch all" definition for the energy stored by the molecules that make up the system. The question is this: if at a certain temperature the molecules cease to store energy because they become essentially motionless, is there a way to obtain that temperature or a lower one?

If there is a way, we have not found it. The absolute zero is the temperature at which molecules are projected to cease to store vibrational energy. Like the limits of the Universe (if they exist) we cannot get there; the molecules always store some energy. Unlike the limits of the Universe, we have gotten mighty close. Some cryogenic experiments have recorded 0.01 K or below (K is degree Kelvin, the unit of temperature in the absolute scale) [2]. So, an absolute temperature exists; and what are the consequences of such a conclusion? Well, this matter is important because we have established that irreversibilities destroy work. We have also studied irreversibilities exclusively as friction, although we have declared them to be capable of assuming many forms. As irreversibilities convert work into heat, and heat flows to lower temperatures, new irreversibilities arise due to heat transfer. Absolute temperature allows us to characterize the irreversibility associated with heat transfer, as seen in what follows.

ENTROPY: DOES IT REALLY EXIST?

To gain a foothold on heat transfer irreversibility that is quite different from friction but that emanates from it, consider entropy. Entropy can be confusing because it is defined based on a process that does not exist, namely a reversible one. Say that a system is exchanging heat through its boundary that is at absolute temperature T. Further, assume that the heat gets from the inside of the system to the boundary reversibly; this would mean that the heat is flowing in such a way that its flow could be reversed without any work. To gain an idea of the difficulty of reversing the flow of heat, a helpful experiment would be to run a kitchen refrigerator unplugged. No matter how long the wait, the refrigerator would never become cooler than the kitchen; as the food spoils, it might get hotter due to bacterial action, but never colder.

So back to this impossibility (Figure 1.11): the heat leaving the system arrives at the boundary reversibly, at a rate dQ_{rev}/dt, and leaves the system at a boundary point

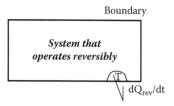

FIGURE 1.11 An internally reversible process.

at T. The rate of entropy change of the system is defined as

$$\frac{dS}{dt} = \frac{1}{T} \cdot \frac{dQ_{rev}}{dt} \tag{1.14}$$

So, entropy rates simply measure how fast heat is being transferred across a system boundary, the proportionality constant being the inverse of the absolute temperature. Entropy is a variable created to provide this characterization of heat flow, involving the absolute temperature. It serves us well because irreversible processes result in net entropy growth, and net entropy growths demand fuel.

ENTROPY GROWS LIKE WEEDS

The fact that entropy can allow the measurement of heat transfer irreversibility is one of the triumphs of Science. Let us call the system in Figure 1.12 a heat source if it loses thermal energy, or a heat sink if it gains it. The heat sources will be at temperatures T_h (hot), whereas the sink is at temperature T_c (cold). We place each system in relative vertical position so that the higher its position, the higher its temperature.

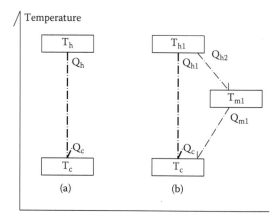

FIGURE 1.12 Heat sources and sinks. Heat flows from hot to cold and entropy increases for one source-sink pair (a) or several (b).

We still assume the systems to be reversible so that we can focus exclusively on the irreversibility generated by heat transfer. In case (a), a large temperature drop exists between the two systems. Heat flows from the h (hot) system to the c (cold) system (another unavoidable fact of life similar to friction). The rate of entropy loss of the hot one, h, is

$$\frac{dS_h}{dt} = \frac{1}{T_h} \cdot \frac{dQ_h}{dt}$$

The rate of entropy gain of the cold system is

$$\frac{dS_c}{dt} = \frac{1}{T_c} \cdot \frac{dQ_c}{dt}$$

Since no heat is lost to any other system, and Q_h is negative whereas Q_c is positive, we have, taking Q as positive,

$$Q = -Q_h = Q_c$$

Thus, replacing Q in both equations for rate of entropy change, we have

$$\frac{dS_h}{dt} = -\frac{1}{T_h} \cdot \frac{dQ}{dt} \qquad \frac{dS_c}{dt} = \frac{1}{T_c} \cdot \frac{dQ}{dt}$$

The net rate of entropy change (there are no irreversibilities in h or c) is then the sum of the change for each system:

$$\frac{dS_{net}}{dt} = \left(\frac{1}{T_c} - \frac{1}{T_h} \right) \cdot \frac{dQ}{dt} \tag{1.15}$$

Equation 1.15 tells us that, when heat flows spontaneously (that is, from hot to cold absolute temperatures), the net entropy increase is positive, because dQ/dt and $(1/T_c - 1/T_h)$ are both greater than zero. It is of interest to establish how the rate of entropy generation varies as the temperature difference varies. For instance, relative to the entropy generation rate at a temperature difference of 1 K (that is, for transfer from 2500 K to 2499 K), the following trace (Figure 1.13) is obtained as T_c decreases.

If more sources/sinks are added (Figure 1.12 (b)), the conclusion is still the same: the entropy generation rate is positive. Also, the greater the entropy generation, the greater the temperature difference for heat transfer.

The skeptical reader may wonder how useful entropy is, regardless of how fast it might grow. We have assumed here ideal systems, but if they were not ideal, all that we would have to do is redraw the boundaries around points where the heat is generated by irreversibilities, and as heat flows out, entropy will decrease inside the new system. However, it will increase to a greater extent in the system receiving the heat. The total entropy generation rate, then, is a measure of how fast heat flows and of the

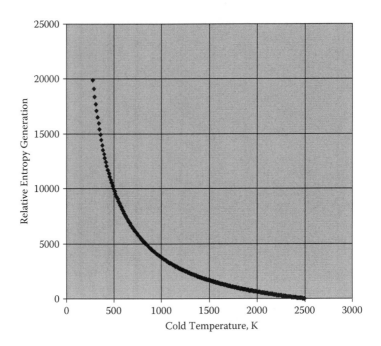

FIGURE 1.13 Entropy generation by heat transfer grows rapidly with the temperature difference (T_h = 2500 K).

magnitude of the temperature difference causing the heat flow. This measure turns out to be a useful way to identify and root out irreversibilities.

Example 1.9 discusses entropy and heat flow.

EXAMPLE 1.9 ENTROPY AND HEAT FLOW

Calculate the net entropy increase for the heat sources/sinks shown in Figure Example 1.9.1 for the temperatures given in each case. The sources and sinks are internally reversible. Entropy is generated due to heat transport across finite temperature differences. Comment on your results.

In both cases,

$$\frac{dQ_{h1}}{dt} = -100 \text{ kW} \qquad \frac{dQ_c}{dt} = -100 \text{ kW}$$

$$\frac{dQ_{h2}}{dt} = -90 \text{ kW} \qquad \frac{dQ_{m1}}{dt} = 60 \text{ kW}$$

Case a

$$Th1a = 1500 \text{ K} \qquad Tm1a = 800 \text{ K} \qquad Tca = 300 \text{ K}$$

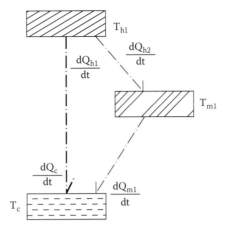

FIGURE EXAMPLE 1.9.1 Heat flowing from a hot source to two other sources.

Case b

$$Th1b = 1500 \text{ K} \qquad Tm1b = 1495 \text{ K} \qquad Tca = 1490 \text{ K}$$

Case a

Systems: We consider each sink/source as a system. We *imagine* that inside each source/sink heat flows reversibly to its boundary.

Assumptions/Conventions: Heat lost is negative, heat gained is positive for each sink/source.

$$Th1a = 1500 \text{ K} \qquad Tm1a = 800 \text{ K} \qquad Tca = 300 \text{ K}$$

The entropy rate of change of source h can be calculated at any instant as

$$\frac{dSh}{dt} = -\frac{\dfrac{dQ_{h1}}{dt} + \dfrac{dQ_{h2}}{dt}}{Th}$$

For case a, we have

$$\left.\frac{dSh}{dt}\right]_a = -\frac{\dfrac{dQ_{h1}}{dt} + \dfrac{dQ_{h2}}{dt}}{Th1a} \qquad \left.\frac{dSh}{dt}\right]_a = -0.127 \,\frac{\text{kW}}{\text{K}}$$

Similarly, for the other two sources,

$$\left.\frac{dSm}{dt}\right]_a = \frac{\dfrac{dQ_{h2}}{dt} - \dfrac{dQ_{m1}}{dt}}{Tm1a} \qquad \left.\frac{dSm}{dt}\right]_a = 0.038 \,\frac{\text{kW}}{\text{K}}$$

$$\left.\frac{dSc}{dt}\right]_a = \frac{\dfrac{dQ_c}{dt} + \dfrac{dQ_{m1}}{dt}}{Tca} \qquad \left.\frac{dSm}{dt}\right]_a = 0.533 \,\frac{\text{kW}}{\text{K}}$$

The net entropy rate of change is then

$$
\left.\frac{dSnet}{dt}\right]_a = \left.\frac{dSh}{dt}\right]_a + \left.\frac{dSm}{dt}\right]_a + \left.\frac{dSc}{dt}\right]_a = 0.444\ \frac{kW}{K}
$$

and the entropy is increasing because heat transfer is occurring with relatively large temperature differences.

Case b

In this case, temperatures are much closer together, so one would get far less entropy generation.

For,

$$
Th1b = 1500\ K \qquad Tm1b = 1495\ K \qquad Tca = 1490\ K
$$

we obtain as before

$$
\left.\frac{dSh}{dt}\right]_b = -\frac{\dfrac{dQ_{h1}}{dt} + \dfrac{dQ_{h2}}{dt}}{Th1b} \qquad \left.\frac{dSh}{dt}\right]_b = -0.126667\ \frac{kW}{K}
$$

Similarly, for the other two sources,

$$
\left.\frac{dSm}{dt}\right]_b = \frac{\dfrac{dQ_{h2}}{dt} - \dfrac{dQ_{m1}}{dt}}{Tm1b} \qquad \left.\frac{dSm}{dt}\right]_b = 0.020067\ \frac{kW}{K}
$$

$$
\left.\frac{dSc}{dt}\right]_a = \frac{\dfrac{dQ_c}{dt} + \dfrac{dQ_{m1}}{dt}}{Tca} \qquad \left.\frac{dSm}{dt}\right]_a = 0.107382\ \frac{kW}{K}
$$

The net entropy rate of change is, for this case,

$$
\left.\frac{dSnet}{dt}\right]_b = \left.\frac{dSh}{dt}\right]_b + \left.\frac{dSm}{dt}\right]_b + \left.\frac{dSc}{dt}\right]_b = 0.00078\ \frac{kW}{K}
$$

Case b generates much less entropy than case a, namely about three orders of magnitude less. In case b, the hot source transfers heat, but the temperature of the sources receiving the heat is still quite high. In a way, it can be said that the *quality* of the heat was preserved, understanding by quality its temperature. Thermodynamics is concerned with temperature, and in assessing quality, the higher the better. Discussion of the matter of quality and work will be postponed until Example 1.10 is presented.

IRREVERSIBILITY AND ENTROPY

Mechanical energy is converted into heat via friction, and the heat thus generated proceeds to flow, generating entropy. Friction is a "catch all" word to denote irreversibility, an ever-present feature in all technical contraptions. In other words, all energy conversion contraptions (the good guys) will produce, because irreversibilities (the bad guys) always arise, an increase in entropy (the impartial referee).

Irreversibilities consume work, transforming it into heat. Hence, devices that require work inputs will require larger inputs the more irreversible they are. Also, devices that produce work will yield smaller work outputs the more irreversible they are. Insightful engineering consists of rooting out capital irreversibilities in the conceptual stage and eliminating as many as possible in the design stage. This is not easy due to a long-standing conflict.

When mechanical work is converted into heat, heat flows to a lower temperature, and the reverse (heat flowing from low to high temperature) has never been observed without additional work expenditures. Hence, heat flow is another key form of irreversibility, which we have already correlated with entropy increases. All real processes entail irreversibilities, approximating reversible behavior is difficult, and obtaining reversibility is impossible. A conflict exists between reversibility and speed. Friction effects increase with speed. Plan a machine to run very slowly, and for the same output, its size increases. Plan a heat exchanger to transfer heat with a minimal temperature difference, and its size becomes impossibly large. We like machines with small footprints that need to run fast to be small. Hence, all our machines will have irreversibility in one degree or another, which is a way to say that they will use more energy than they should. It gets even worse. A large segment of thermomechanical machines will not reach an efficiency of one even if we could make them perfectly reversible.

CONVERSIONS—CYCLICAL AND DIRECT

CONVERSION OF HEAT INTO WORK

Presumably, heat flowed gingerly from hot to cold temperatures for eons, and then along came humans. Because of the species' intelligence and desire for progress, work—mechanical work—needs to be done. Plowing requires mechanical work; so does placing a man on the moon. Whereas fire was harnessed early for security, cooking, and warmth, it took some time for humans to produce work from fire. It was relatively recently (1700s) that fire was applied to generate work with heat engines. Conceptually, a heat engine receives heat from a hot source at a certain rate and converts as much heat as possible into power. The key question is how much is "as much as possible." This question occupied the minds of many bright individuals (Carnot, Kelvin, and Clausius being the most prominent ones) in the 18th century. Simpler minds probably saw no reason to speculate on the question at hand, and perhaps they deemed the entire conversion of thermal energy into work as a feasible process. However, this question is ever more pressing as energy supplies become more and more strained.

In principle, an engine could operate as illustrated in Figure 1.14(a). The engine is assumed to operate in cyclic fashion, namely, to return, after converting heat into work, to its original state. This is key; if the engine does not change after a cycle, then, for any number of complete cycles, the machine will be unchanged. If the machine is unchanged, it is immune to irreversibilities, and its entropy should remain the same.

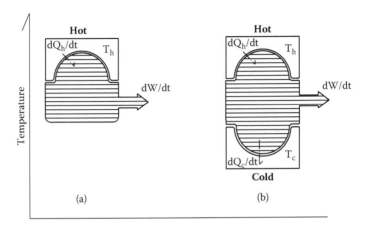

FIGURE 1.14 Heat engines and temperature, (a) single heat source, (b) heat source, engine and heat sink.

Using the subscript e for the engine, we have, for its entropy change,

$$\frac{dS_e}{dt} = 0$$

In Figure 1.14(a), we assume that our machine is reversible and that it receives heat reversibly, that is, without a temperature difference between the hot source and the machine. The heat source has lost heat (which supposedly has all gone into power, dW/dt), and hence, using the subscript h for the heat source, we have

$$\frac{dS_h}{dt} = \frac{1}{T_h} \cdot \frac{dQ_h}{dt} < 0$$

which brings us to the conclusion that the net entropy change, namely,

$$\frac{dS_{net}}{dt} = \frac{dS_e}{dt} + \frac{dS_h}{dt} < 0$$

is less than zero. Since as of today no recorded net entropy decrease has been observed, the foregoing equation indicates an impossibility. This is a reason for alarm, or at least a reason not to invest large funds in developing such an engine. Irreversibilities are the normal state of affairs, and they always generate entropy.

We are, then, obliged to take a second look at our engine. In an optimistic vein, we assume it to be reversible still. Then, for this machine to actually work, the entropy generation rate has to be zero if the heat transfer to and from the machine is effected reversibly, as in Figure 1.14(b). This is a Carnot engine. In this case, the net entropy change (there being no irreversibilities, just a perfect machine humming away), will be zero, which brings us to a most interesting conclusion (taking the absolute value of all heat exchanged and using minus signs to adjust for heat lost or gained) we have, for this case

$$\frac{dS_{net}}{dt} = \frac{dS_e}{dt} + \frac{dS_h}{dt} + \frac{dS_c}{dt} = \frac{dS_e}{dt} - \frac{1}{T_h} \cdot \frac{dQ_h}{dt} + \frac{1}{T_c} \cdot \frac{dQ_c}{dt} = 0 \qquad (1.16)$$

For a complete number of cycles, the energy change of our engine is zero, and we can write from the First Law

$$\frac{dQ_h}{dt} - \frac{dQ_c}{dt} = \frac{dW}{dt} \qquad (1.17)$$

If we define the efficiency of this engine as the ratio of heat input to heat output, we can arrive at an interesting conclusion. From Equation 1.16, we get

$$dQ_c = \frac{T_c}{T_h} \cdot dQ_h$$

which, combined with Equation 1.17, indicates that

$$dW = \left(1 - \frac{T_c}{T_h}\right) \cdot dQ_h$$

and the Carnot engine efficiency becomes

$$\eta_{Ca} = \frac{dW}{dQ_h} = 1 - \frac{T_c}{T_h} \qquad (1.18)$$

Hence, this most unique conceptual engine requires a heat sink and a heat source to operate if the principle of zero entropy generation for a reversible process is to be heeded. Also, its efficiency is the maximum theoretically possible since the engine is reversible and the heat transfer occurs reversibly as well. The efficiency is a function of the two temperatures needed to generate power—that of the hot source and that of the cold source.

This is a remarkable result: engines with additional irreversibilities, arising from the generation, transmission, and rejection of heat, or due to friction between moving components, or in general due to the conversion of mechanical energy into heat, will have lower efficiencies than those given by Equation 1.18. So far in the history of technology, no cyclical engine has converted heat into work at efficiencies even approaching Carnot's. We show the Carnot efficiencies and the actual efficiency ranges (Figure 1.15) for a number of commonly used engines to highlight this conclusion. The Otto cycles are implemented predominantly in automobile engines; Diesel cycles power cars and many trucks and heavy equipment; Brayton cycles are implemented in aircraft engines and in power generation; and Rankine cycles are also used for power generation. What all those cycles have in common is that they are driven by a hot source, and that the hot source is generally the flame and combustion products of fossil fuels. In Rankine cycles, the hot source is sometimes nuclear

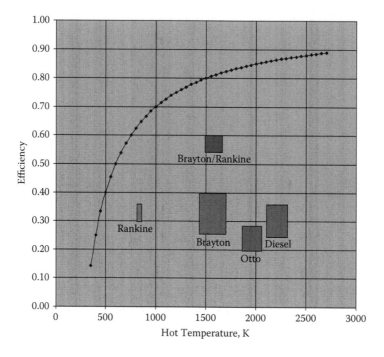

FIGURE 1.15 Range of efficiency versus peak firing temperature for Carnot and practical cycles, for a sink temperature of 300 K.

fission. None of the heat sources provides a constant temperature or reversible heat transmission or generation. The range of peak temperatures and the efficiencies adopted are approximate.

For the real cycles, entropy generation is always positive. Irreversibility spawns heat that ends up in the atmosphere. If the engine efficiency is η_e, we have

$$\eta_e = \frac{\dfrac{dW}{dt}}{\dfrac{dQ_h}{dt}}$$

and the First Law applied to the engine (Figure 1.14) gives

$$\frac{dQ_h}{dt} - \frac{dQ_{atm}}{dt} = \frac{dW}{dt}$$

where the heat and work flows are taken as positive, with the sign reflecting direction given in the equation. From both equations, one can obtain

$$\frac{dQ_{atm}}{dt} = \frac{dW}{dt} \cdot \left(\frac{1}{\eta_e} - 1 \right)$$

$$\frac{dQ_h}{dt} = \frac{dW}{dt} \cdot \frac{1}{\eta_e}$$

Since the total entropy generation is that of the atmosphere receiving the heat minus that of the hot source delivering the heat, using both foregoing equations we obtain

$$\frac{dS_{gen}}{dt} = \frac{1}{T_{atm}} \frac{dQ_{atm}}{dt} - \frac{1}{T_h} \cdot \frac{dQ_h}{dt} = \frac{dW}{dt} \cdot \left[\frac{1}{\eta_e \cdot T_{atm}} \cdot \left(1 - \frac{T_{atm}}{T_h} \right) - \frac{1}{T_{atm}} \right] > 0 \qquad (1.19)$$

The entropy generation rate only goes to zero if the efficiency approaches Carnot, deemed something of an impossibility. In this case, Equation 1.19 becomes a null identity:

$$\frac{dS_{gen,rev}}{dt} = \frac{dW}{dt} \cdot \left[\frac{1}{\left(1 - \frac{T_{atm}}{T_h} \right) \cdot T_{atm}} \cdot \left(1 - \frac{T_{atm}}{T_h} \right) - \frac{1}{T_{atm}} \right] = \frac{dW}{dt} \cdot \left[\frac{1}{T_{atm}} - \frac{1}{T_{atm}} \right] = 0$$

Example 1.10 discusses heat and its potential to become work.

EXAMPLE 1.10 POTENTIAL OF HEAT FOR CONVERSION INTO WORK

Consider two Carnot engines (Figure Example 1.10.1) working with the same sink and receiving the same input heat rate but at different temperatures. Each machine has a different heat source.

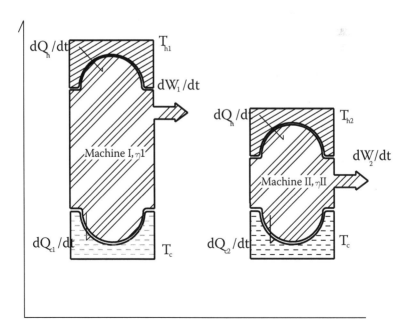

FIGURE EXAMPLE 1.10.1 Heat engines.

Calculate

a. Which machine will convert a greater portion of the heat input into work
b. Explain why one machine (they are both reversible and receive heat at the same rate!) has a greater potential for conversion than the other.

$$Th1 = 2000 \text{ K} \qquad Th2 = 1300 \text{ K} \qquad Tc = 300 \text{ K}$$

$$\frac{dQh}{dt} = 500 \text{ kW}$$

Solution

System: Each machine works separately. Each machine inputs heat at the same rate, but each operates at different efficiencies. For each machine (1 being the one on the left of the diagram), we have

$$\eta1 = 1 - \frac{Tc}{Th1} \qquad \eta2 = 1 - \frac{Tc}{Th2}$$

$$\eta1 = 0.85 \qquad \eta2 = 0.77$$

The power output then for machine 1 is

$$\left.\frac{dW}{dt}\right]_1 = \eta1 \cdot \frac{dQh}{dt} = 4.25 \cdot 10^5 \text{ W}$$

and for machine 2

$$\left.\frac{dW}{dt}\right]_2 = \eta2 \cdot \frac{dQh}{dt} = 3.85 \cdot 10^5 \text{ W}$$

Hence, machine 1 has a greater power output. One way to look at this result is to assign a greater work potential to the higher temperature heat input. In a reversible world (and in the real one too), higher temperature heat has a greater potential to be transformed into work. All the heat at the sink temperature has no capacity to do work.

For machine 1
$$\frac{dQc1}{dt} = \frac{dQh}{dt} - \left.\frac{dW}{dt}\right]_1 = 75000 \text{ W}$$

For machine 2
$$\frac{dQc2}{dt} = \frac{dQh}{dt} - \left.\frac{dW}{dt}\right]_2 = 115400 \text{ W}$$

The machine operating with the lower temperature heat source generates waste heat at a much higher rate than the one with the higher temperature heat source.

CONVERSION OF WORK INTO WORK

Via the Second law, we have established that processes exhibiting irreversibilities will generate entropy at a rate proportional to the irreversibility associated with the process. Consider for instance a shaft that receives power, transmits it, and delivers power at the other end. Bearings will heat up and unsteady loads will deform the shaft, with some associated dissipation. Less power will be delivered. The ratio of output to input power is the mechanical efficiency η_m. Typically, this efficiency is large. However, because of the work destroyed, a higher power input will be required for a given output, and more fuel will be spent. Unlike the case of the heat engine, the Second law does not predict a limit to η_m, which clearly cannot exceed one because of the First law.

Friction effects are assessed via entropy generation. All the heat generated by our shaft will end up, via oil coolers or directly, in the environment. The entropy generation rate in the environment is simply

$$\frac{dS_{atm}}{dt} = \frac{1}{T_{atm}} \cdot \frac{dQ}{dt}$$

Since mechanical efficiency is related to heat production as follows, we have (Figure Example 1.11.1)

$$\eta_m = \frac{\dfrac{dW}{dt}\bigg]_{out}}{\dfrac{dW}{dt}\bigg]_{in}} = \frac{\dfrac{dW}{dt}\bigg]_{in} - \dfrac{dQ}{dt}}{\dfrac{dW}{dt}\bigg]_{in}}$$

Combining both the foregoing equations, we obtain

$$\frac{dS_{atm}}{dt} = \frac{dS_{gen}}{dt} = \frac{(1-\eta_m) \cdot \dfrac{dW}{dt}\bigg]_{in}}{T_{atm}} \tag{1.20}$$

Hence, all real energy conversion devices, mechanical to mechanical, will destroy work that, when dissipated to the atmosphere as heat, generates entropy. This is the case for a shaft, or a gearbox, or, with suitable modification of the definition of efficiency (mechanical power out to electric power in), for an electric motor.

Example 1.11 discusses power transmission and entropy growth.

EXAMPLE 1.11 POWER TRANSMISSION AND IRREVERSIBILITY

According to Smil,[*] measurements on a well-preserved Dutch windmill from the 18th century indicated transmission input and output power as follows. Calculate the efficiency and entropy generation rate of this transmission if all the heat generated by friction ultimately ends up in the environment at T_{atm}. The transmission can be considered to be at T_{trans}.

$$\frac{dW}{dt}\bigg]_{in} = 30 \text{ kW} \qquad \frac{dW}{dt}\bigg]_{out} = 12 \text{ kW}$$

$$T_{atm} = 300 \text{ K} \qquad T_{trans} = 340 \text{ K}$$

---

[*] Smil, V. 1994, *Energy in World History*. Westview Press, Boulder, Colorado.

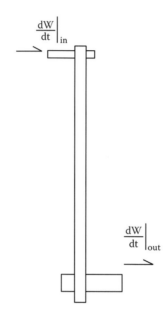

FIGURE EXAMPLE 1.11.1 Transmission schematic.

Solution

System: The transmission (Figure Example 1.11.1).
   The efficiency is simply

$$\eta_m = \frac{\dfrac{dW}{dt}\bigg]_{out}}{\dfrac{dW}{dt}\bigg]_{in}} = 0.4$$

The First Law indicates that, in steady state, heat is generated at a rate

$$\frac{dQ}{dt} = \frac{dW}{dt}\bigg]_{in} - \frac{dW}{dt}\bigg]_{out} = 18000\ \text{W}$$

The entropy generated is the difference between the entropy lost by the transmission and that gained by the environment:

$$\frac{dS}{dt} = \frac{dQ}{dt} \cdot \left(\frac{1}{T_{atm}} - \frac{1}{T_{trans}} \right) = 7.06\ \frac{\text{W}}{\text{K}}$$

Hence, a low-efficiency mechanical device is not exempt from irreversibilities, and the windmill power output is reduced. The power reduction appears as heat dissipated to the environment, and it has no longer any capacity to do work.

CONVERSION OF WORK INTO HEAT

This conversion is an easy one, for just rubbing your hand against the table will heat up both. The mechanical work from your muscles goes into heat handily, so to speak. Even though heat engines, including the best and the mythical, cannot convert all heat into work, work into heat is such a simple conversion that anyone can do it with 100% efficiency. Consider an electrical heater. It receives perfectly good electrical power and converts it into heat that flows. This work could have been used to run an electric motor with a useful effect, but once it becomes heat, only a portion can be turned back into work. Once dissipated into the atmosphere, the work, and the fuel used to generate it, are gone for all the purposes of a human time scale. The entropy generation rate is, in this case,

$$\frac{dS_{gen}}{dt} = \frac{\dfrac{dW}{dt}}{T_{atm}} \tag{1.21}$$

CONVERSION OF CHEMICAL ENERGY INTO WORK

Direct conversion of chemical to electrical energy can be done by fuel cells or batteries, or other devices starting with the same fuel (Figure 1.16). Consider a device that receives hydrogen and oxygen, releasing water and power and perhaps thermal energy either contained in the exhaust stream or as reject heat. The practical efficiency of such an engine, η_{cc}, is simply the power out divided by the heating value of the fuel (HV) times its mass flow rate ($\dot{m}_f$). We then have for efficiency

$$\eta_{cc} = \frac{\dfrac{dW}{dt}}{\dot{m}_f HV_f}$$

and an energy balance around the device gives

$$\dot{m}_f \cdot HV_f - \frac{dQ_c}{dt} = \frac{dW}{dt}$$

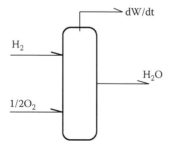

FIGURE 1.16 A chemical engine produces power from a chemical reaction.

where Q_c represents the thermal energy leaving the system. Introducing dS_{cc}/dt as the entropy change associated with the change from reactants to products and other internal irreversibilities, and noting that the input is either converted into power or into a heat rate rejected to the atmosphere as $Q_{atm} = Q_c$, we get that the entropy variation must exceed zero, or

$$\frac{dS_{gen}}{dt} = \frac{dS_{cc}}{dt} + \frac{1}{T_{atm}} \cdot \frac{dQ_{atm}}{dt} = \frac{dS_{cc}}{dt} + \frac{\frac{dW}{dt}}{T_{atm}} \cdot \left(\frac{1}{\eta_{cc}} - 1 \right) > 0 \qquad (1.22)$$

The Second Law does not impose a maximum efficiency on these devices, but as irreversibilities always arise, it is clear that overall entropy generation must be positive. As shown by Equation 1.22, dS_{cc} could be negative if its value is overridden by the power output term. The smaller η_{cc} is for a given power output, the smaller dS_{cc} can be. If an additional heat input is considered from an additional source at a higher temperature than that of the cc, the definition of efficiency changes to

$$\eta_{cc} = \frac{\frac{dW}{dt}}{\dot{m}_f \cdot HV_f + Q_h}$$

where Q_h stands for the heat from the hot source. Then the overall entropy generation rate becomes

$$\frac{dS_{gen}}{dt} = -\frac{Q_h}{T_h} + \frac{dS_{cc}}{dt} + \frac{\frac{dW}{dt}}{T_{atm}} \cdot \left(\frac{1}{\eta_{cc}} - 1 \right) > 0 \qquad (1.23)$$

with the surprising conclusion that η_{cc} will have to decrease below the value of Equation 1.22 (all other variables being constant) for the entropy generation rate to be zero (reversible yet impossible case) or the entropy generation rate to be greater than zero (actual case).

Example 1.12 discusses chemical engine and entropy changes.

EXAMPLE 1.12 CHEMICAL ENGINE

Consider a 1-kW chemical engine, with the efficiency as follows. The atmosphere is at the given temperature. Calculate the minimum entropy generation rate dS_{cc}/dt associated with this process.

$$\frac{dW}{dt} = P = 1 \, \text{kW} \qquad T_{atm} = 300 \, \text{K} \qquad \eta_{cc} = 0.6$$

Solution

From Equation 1.22,

$$\frac{dS_{gen}}{dt} = \frac{dS_{cc}}{dt} + \frac{dQ_{atm}}{dt} = \frac{dS_{cc}}{dt} + \frac{P}{T_{atm}} \cdot \left(\frac{1}{\eta_{cc}} - 1 \right) > 0$$

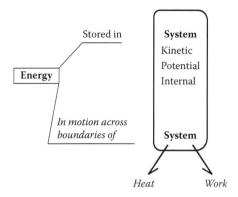

FIGURE 1.17 Forms of energy and systems.

The minimum entropy generation rate is

$$\frac{dS_{cc}}{dt} \geq -\frac{P}{T_{atm}} \cdot \left(\frac{1}{\eta_{cc}} - 1\right) = -2.22 \, \frac{W}{K}$$

Hence, operating at this level of efficiency, the entropy generation rate of the chemical engine could be, quite remarkably, negative.

SUMMARY

Energy and systems go together. Energy can be stored in a system or can be in transit across the system boundaries. Graphically, this is shown in Figure 1.17. Whereas energy is conserved (First Law), heat and work are not entirely equivalent as the equality (Figure 1.18) would lead us to surmise.

The conversion of energy is constrained (Second Law). In any process, cyclical or not, some of the mechanical energy will become heat even if we do not intend it to. Even worse, once we have heat, not even reversible machines can convert it all back into work. Work or heat, when they are sent over distribution networks (such as electrical energy via high voltage lines or thermal energy via a furnace and air distribution systems) are subject to irreversible losses, and the end product does not embody as much energy as the input energy. In Figure 1.19, dotted arrows show constrained conversions (cannot be complete), and solid arrows

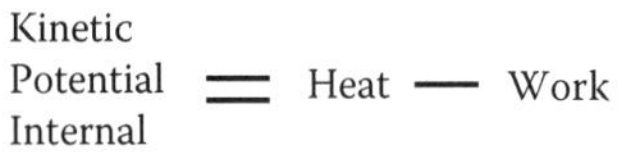

$$\begin{array}{c} \text{Kinetic} \\ \text{Potential} \\ \text{Internal} \end{array} \quad = \quad \text{Heat} \quad - \quad \text{Work}$$

FIGURE 1.18. First Law: this law fails to show that heat and work are not entirely equivalent.

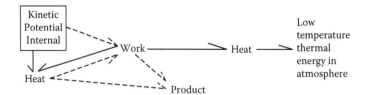

FIGURE 1.19 The Second Law indicates that some conversions can be complete but that mechanical energy eventually evolves into heat.

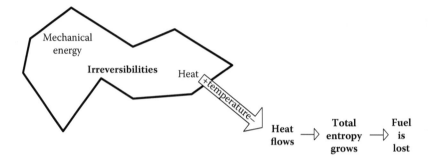

FIGURE 1.20 Irreversibilities increase fuel consumption.

show conversions that can be conceptually complete, although most of them will be subject to irreversibility.

Entropy is a gauge of irreversibility, which demands additional fuel consumption for delivering a certain amount of utility, be it mechanical or thermal energy. Figure 1.20 illustrates the interrelationship between mechanical energy and lost fuel due to irreversibilities.

REFERENCES

1. Larousse 2003. *El Pequeño Larousse Ilustrado*, Ed Larousse, Mexico.
2. Lynds, B. T. About Temperature, http://eo.ucar.edu/skymath/tmp2.html.

2 The Equations for Transient Phenomena

It is all in the timing...
(Popular aphorism)

Mass, momentum, and energy transfer in transient regimes follow at every instant the laws of Chapter 1. Changes in energy (or enthalpy) reflect changes of what is stored in matter. At every instant that real processes unfold, total entropy increases. The entropy increase signals a decrease in the quality of energy, the loss of which cannot be made up in a short time scale. This loss translates into a decrease of the capacity to perform work. The transient form of the laws and entropy-generation equations and elementary forms of relevant transport equations are introduced here. Simulation programs are included on the CD.

STATE, PROPERTIES, AND PROCESS

Temperature defines a state of a system, that is, a system at 400 K is hotter than one at 300 K. Temperature is a property of the system. Pressure and density also define the state of a system. Energy in any of its stored forms is likewise useful for defining the state of a system. The state, then, is defined by a set of conditions that completely determine what we need to know about the system. When a system goes from one state to another, it follows a path. This path may be well defined (we may know all the properties at all times), or it may be poorly defined (properties within the system assume different values at the same time in different parts of the system). The transformation from one state to the next is a process.

There is no such thing as a reversible process, in that neither system nor surroundings can be returned to their initial state without work expenditure. Some processes are internally reversible, meaning that, if we know the properties that define the state at all times during a process, we can make the system follow the reverse path, although that is likely to require work. Usually, two properties are required to know the state of a system, for example, pressure and temperature, as shown in the following text (Figure 2.1).

Some properties depend on the mass of the system (they are called extensive properties), whereas others do not (they are called intensive properties). Energy is an extensive property. Double the mass of a system, and the kinetic energy doubles. The system temperature, though, does not depend on the amount of mass present. Temperature is then an intensive property. The specific energy of a system, namely its energy divided by its mass, is also a property, just like pressure and temperature. Hence,

$$e = \frac{E}{m} = \frac{V^2}{2} + g \cdot z + u \tag{2.1}$$

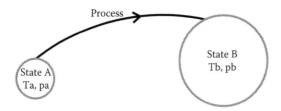

FIGURE 2.1 System process from State A to State B.

Specific energy, e, has a value independent of the mass, and hence e is intensive. Entropy is also a property of a mass m:

$$s = \frac{S}{m} \tag{2.2}$$

Therefore, mass can embody energy or entropy as a property. The process whereby the mass acquired entropy or energy is immaterial. An infinite number of paths could have been followed to arrive at any state: one can arrive in Rome from many different points via different transportation means: walking, bicycle, car, bus, plane, train; the important thing is to be there, not the method of arrival or the path followed.

ENTHALPY

If three functions are properties, then it follows that any algebraic combination of them will also be a property. Of all the possible combinations of properties, enthalpy conveys deep physical meaning, and hence we consider it separately. Enthalpy is defined as

$$h = u + \frac{p}{\rho} = u + p \cdot v \tag{2.3}$$

where v stands for specific volume, that is, the inverse of density. Enthalpy then measures the energy content u of the unit mass, plus the product of the pressure multiplied by the volume. The reader may recall from Chapter 1, Equation 1.7, that differential work is the product of pressure multiplied by a differential of volume. Hence, enthalpy contains the work done by a constant pressure p when a volume v (the volume per unit mass) is swept.

Example 2.1 shows specific energy and states.

EXAMPLE 2.1 SPECIFIC ENERGY AND STATES

Consider a closed system formed by air, considered a perfect gas. With reference to Figure 2.1, the gas evolves from state A to state B:

$$p_a = 100 \text{ kPa} \qquad p_b = 200 \text{ kPa} \qquad c_v = 725 \text{ J/kg} \cdot \text{K}$$

$$T_a = 300 \text{ K} \qquad T_b = 420 \text{ K} \qquad m = 600 \text{ kg}$$

a. Determine the change in internal energy (specific and total) from A to B.
b. The gas has been compressed. Is it possible to calculate the work of compression with the information given?

Solution

a. For the change between A and B, perfect gases have internal energies that are independent of pressure. The specific change is simply given by

$$\Delta u = c_v \cdot (T_b - T_a) = 725 \cdot (420 - 300) = 87000 \text{ J}$$

The total change is given by

$$\Delta U = m \cdot c_v \cdot (T_b - T_a) = 600 \cdot 725 \cdot (420 - 300) = 52.2 \cdot 10^6 \text{ J}$$

b. In the absence of more information, the work and heat exchanged to drive the gas from A to B cannot be calculated.

Example 2.2 shows specific enthalpy changes.

EXAMPLE 2.2 SPECIFIC ENTHALPY CHANGE

Consider a closed system formed by air, considered a perfect gas. With reference to Figure 2.1, the gas evolves from state A to state B:

$$p_a = 100 \text{ kPa} \qquad p_b = 200 \text{ kPa} \qquad c_p = 1000 \text{ J/kg} \cdot \text{K}$$

$$T_a = 300 \text{ K} \qquad T_b = 420 \text{ K} \qquad m = 600 \text{ kg}$$

a. Determine the change in enthalpy (specific and total) from A to B.

Solution

a. For the change between A and B, perfect gases have enthalpies that are independent of pressure. The specific change is simply given by

$$\Delta h = c_p \cdot (T_b - T_a) = 1000 \cdot (420 - 300) = 120000 \text{ J}$$

The total change is given by

$$\Delta H = m \cdot \Delta h = 600 \cdot 120000 = 72 \cdot 10^6 \text{ J}$$

REYNOLD'S TRANSPORT THEOREM

Somewhat like credits, debits, and balance in a checking account, energy gains, losses, and energy storage rates must balance. Somewhat similar to the acquisitive power of money, the quality of energy will never balance because it changes with time. The acquisitive power of money can improve (the currency becomes stronger relative to the goods of interest) or decrease (as anyone on a fixed income can attest, a dollar today buys less real estate or gasoline than 5 or 10 years ago). Unlike currency,

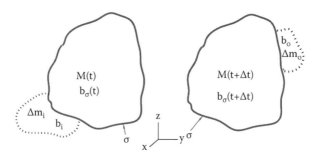

FIGURE 2.2 Control volume at two time instants separated by Δt.

the value of which can fluctuate, the quality of energy always decreases. We present here the Reynold's Transport Theorem, a general accounting method whereby we can keep track of energy balances and also decreases in its quality.

Consider a control volume; this is a portion of space with boundaries permeable to mass, heat, and work. The mass inside this volume is then changing with time, and our system is no longer a fixed amount of mass but a volume such as the one surrounding the windowpane in Figure 1.1 in Chapter 1. It is of fundamental importance to see that properties apply to a mass, whereas the control volume simply isolates a volume (not mass) from space. In Figure 2.2, we show the volume σ, and an amount of mass Δm_i entering at time t. As explained earlier, mass has intensive properties, such as energy e or temperature T, which we label generically as b_i. When a finite time interval Δt has elapsed, Δm_i is now inside σ, and another amount of mass Δm_o has emerged from σ, with properties b_o. Inside the control volume, a certain amount of mass $M(t)$ is present, with properties $b_\sigma(t)$. In addition to time, an x, y, z coordinate system is adopted.

The extensive property corresponding to b is B, defined as

$$B = m \cdot b \tag{2.4}$$

The Laws of Chapter 1 apply to a fixed amount of mass. Our intent is to relate those laws to the control volume σ, which encloses some mass but also receives and discharges mass across its boundaries. In order to find a way to apply the Laws to σ, let us now focus on a fixed amount of mass **sm**. The mass under consideration is moving in space, and at time t, it comprises Δm_i plus $M(t)$ (Figure 2.2), and after Δt has elapsed, the mass **sm** has moved, and we identify it as Δm_o plus $M(t + \Delta t)$. The change ΔB of the property B during the time interval is

$$B = B(t + \Delta t) - B(t) = [b_o \cdot \Delta m_o + M(t + \Delta t) \cdot b_\sigma(t + \Delta t) - (b_i \cdot \Delta m_i + M(t) \cdot b_\sigma(t))].$$

The rate of change of B is obtained dividing the foregoing by the time interval:

$$\frac{\Delta B}{\Delta t} = b_o \cdot \frac{\Delta m_o}{\Delta t} - b_i \cdot \frac{\Delta m_o}{\Delta t} + \left[\frac{M(t + \Delta t) \cdot b_\sigma(t + \Delta t) - M(t) \cdot b_\sigma(t)}{\Delta t} \right] \tag{2.5}$$

The extensive property B changes with time and position because the mass under study is in motion as it traverses the control volume. Hence, in the limit, when $\Delta t \to 0$, each term of the foregoing equation takes on a special meaning. From left to right, we get Rate of change of B present in the closed mass **sm**:

$$\lim_{\Delta t \to 0} \frac{\Delta B}{\Delta t} = \frac{DB}{Dt}$$

Mass flow rate leaving the CV:

$$\lim_{\Delta t \to 0} \frac{\Delta m_o}{\Delta t} = \frac{dm_o}{dt} = \dot{m}_o$$

Mass flow rate entering the CV:

$$\lim_{\Delta t \to 0} \frac{\Delta m_i}{\Delta t} = \frac{dm_i}{dt} = \dot{m}_i$$

Rate of change of total B inside CV:

$$\lim_{\Delta t \to 0} \frac{M(t+\Delta t) \cdot b_\sigma(t+\Delta t) - M(t) \cdot b_\sigma(t)}{\Delta t} = \frac{d(M \cdot b_\sigma)}{dt} = \frac{dB_\sigma}{dt}$$

Whereas the second and third derivatives in the foregoing equations are of ready interpretation, the first and last ones deserve comment. The CV contains mass, and the mass inside the CV is changing. Concerning the last derivative, at any instant, the mass inside the CV has a certain total value of property B. The rate of change of this value inside the CV is dB_σ/dt. The first derivative, DB/Dt, is quite different from dB_σ/dt. Whereas the latter is concerned with changes of B inside σ, the former is the rate of change of the B property in **sm**, which is a fixed mass in motion. Thus, DB/Dt focuses on how B changes within **sm**. When we replace the derivatives in Equation 2.5 after taking the limit for $\Delta t \to 0$, we obtain

$$\frac{DB}{Dt} = \frac{dB_\sigma}{dt} + \dot{m}_o \cdot b_o - \dot{m}_i \cdot b_i \tag{2.6}$$

which simply states that the rate of change of B for a closed system moving with reference to a CV σ equals the rate of change of B inside σ, plus the rate at which a stream leaving σ carries b out, minus the rate at which a stream entering σ brings b in.

The importance of Equation 2.6 hinges on the fact that it relates the rate *of change of an extensive property of a closed (moving) system to the rate of change inside the control volume.* Hence, all the laws for closed systems can be applied to the LHS to obtain the corresponding form of the law for a control volume in the RHS. In what follows, we use this equation to derive useful forms of the laws for control volumes, needed for transient simulations of many thermal systems. Further insight may be derived from Figure 2.3: The LHS of Equation 2.6 applies to the moving closed system of mass **sm**, whereas the RHS applies to CV σ, with permeable boundaries. As **sm** moves through σ, the laws for a closed system apply to the LHS of Equation 2.6, whereas the exchanges across the σ boundaries are described by the RHS at any instant in time.

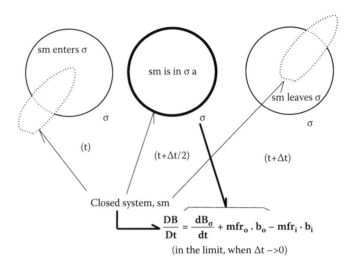

$$\frac{DB}{Dt} = \frac{dB_\sigma}{dt} + mfr_o \cdot b_o - mfr_i \cdot b_i$$

(in the limit, when $\Delta t \rightarrow 0$)

FIGURE 2.3 A pictorial depiction of a closed system traveling through a CV.

CONSERVATION OF MASS

Except for nuclear reactions not considered here, mass is conserved. It cannot be created or destroyed. Consider, then, mass as our property. For our closed-system mass **sm**, we obtain for B and b (Equation 2.4)

$$b = 1 \quad B = sm$$

Because mass is conserved, **sm** does not change with time. Our closed system law is

$$\frac{D(sm)}{Dt} = \frac{DB}{Dt} = 0$$

When we apply Equation 2.6 to a control volume, we find, replacing 1 for b and (**sm**) for B,

$$\frac{D(sm)}{Dt} = 0 = \frac{dM(t)}{dt}. \tag{2.7}$$

Or, simply put, from Equation 2.7, the rate of mass change in σ equals the inlet mass flow rate minus the outlet flow rate:

$$\frac{dM(t)}{dt} = \dot{m}_i - \dot{m}_o \tag{2.7a}$$

Example 2.3 (VisSim) shows how to integrate the foregoing Equation 2.7 via analog programming.

EXAMPLE 2.3 EMPTYING A TANK [VISSIM VERSION ON CD]

Those with no experience in analog programming may want to read Chapter 9 of this book first.

A tank containing an initial mass of liquid M_o, receives at time zero a stream of water with mass flow rate of time dependence:

$$\dot{m}_i(t) = 1 \cdot t \text{ kg/s.}$$

After a time interval Δt has elapsed, water is removed from the tank at a rate

$$\dot{m}_o(t) = 5 \cdot t \text{ kg/s}$$

$$\Delta t = 3 \cdot s$$

$$M_o = 100 \text{ kg}$$

Determine via analog simulation

a. The time for the tank to become empty.

Solution

System: We take as CV the tank.

Application of Equation 2.7a for mass conservation yields

$$\frac{dM(t)}{dt} = \dot{m}_i(t) - \dot{m}_o(t) \tag{2.3.1}$$

The boundary condition is given at time zero as

$$M_o = 100 \text{ kg} \tag{2.3.2}$$

Please follow this simulation in either VisSim© or Simulink©. (Both versions are provided on the CD, but only VisSim is explained here.) The inlet flow rate is a ramp block with a slope of 1. Block properties are set by right-clicking on the block and following instructions on the screen.

Inlet Flow, kg/s

Similarly, the exit flow is set with a delay of 3 s and with a slope of 5. A step function activates the exit flow at 3 s by changing its value from 0 to 1.

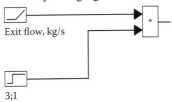

Exit flow, kg/s

3;1

Plotting both flows by selecting the plot block and connecting the output from each block gives the following graph (Figure Example 2.3.1).

Since the exit flow exceeds the inlet at 3.8 s, it is clear that the tank will be depleted. We must answer when this will happen. If we take the difference between the inlet and exit flow, we obtain $dM(t)/dt$, the rate of change of the mass inside the tank. Integrating this rate yields the mass inside the tank at any instant:

$$M(t) = \int_0^t \frac{dM(t)}{dt} \cdot dt = \int_0^t [\dot{m}_i(t) - \dot{m}_o(t)] \cdot dt$$

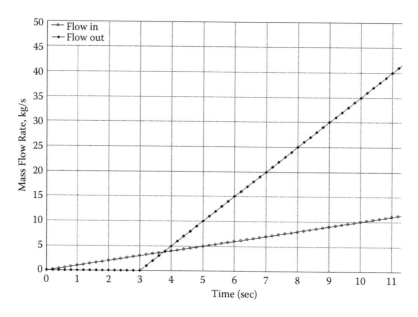

FIGURE EXAMPLE 2.3.1 Flows versus time.

Hence, by taking the difference between the mass flow rates, and integrating using the initial condition, it is possible to ascertain when the tank will be empty. A summing junction is used to calculate the difference in the flows. This signal from the summing block is the input to an integrator block, with an initial condition of 100 kg. The program looks like this:

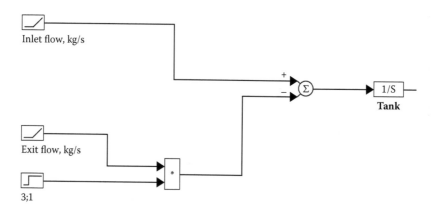

The trace of mass versus time is shown in Figure Example 2.3.2.

The mass in the tank initially increases but, as the larger exit flow dominates, the tank is emptied in 11 s, when the mass equals zero.

Example 2.4 (VisSim) shows mass ejected by a rocket engine.

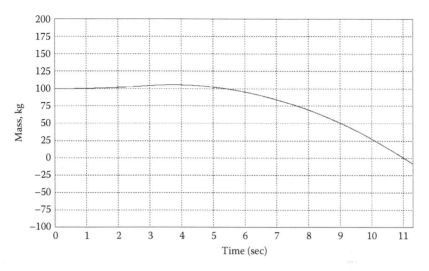

FIGURE EXAMPLE 2.3.2 Mass in the tank versus time.

EXAMPLE 2.4 ROCKET MASS [VISSIM]

A rocket fires its engine, expelling mass according to the following schedule:

From $t_o = 0$ s to $t_1 = 5.56$ s, $mfr1 = 0.1 \cdot t + 0.01 \cdot e^{0.1t}$

For $t > 5.56$ s, $mfr2 = e^{-0.1t}$

Calculate the total mass ejected in 50 s.

Solution

The equation to model is $\dfrac{dM(t)}{dt} = \dot{m}_i - \dot{m}_o$, with $\dot{m}_i$ equal to zero. Each of the exit functions $mfr1$ and $mfr2$ is easy to build, and this is what they look like

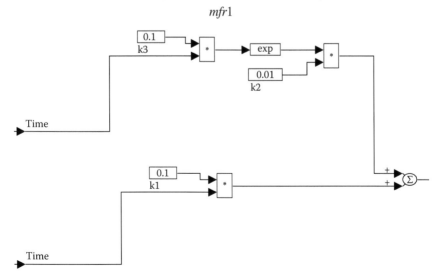

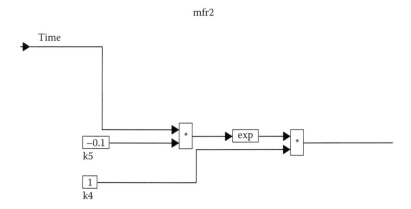

To obtain the corresponding function at each point in time, we multiply each by a factor that is one only when the function is active. One way to do this is as follows using two Boolean blocks.

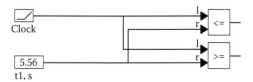

When the time is less than 5.56 s, the upper block is one and the lower block is zero. When the time exceeds 5.56 s, the upper block switches to zero and the lower to one. Hence, each of the foregoing functions is multiplied by the output of the corresponding Boolean block. If each function is transformed into a compound block, (each of the three block diagrams is made into a compound block, that is, a block that contains all the individual blocks of the function), we have, then, the following program.

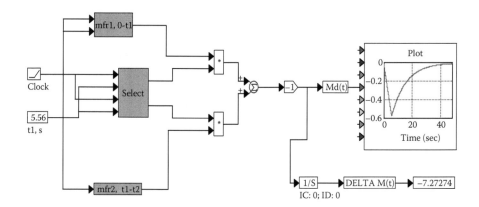

FIGURE EXAMPLE 2.4.1 Rocket experiences thrust as mass is ejected.

The integral of the mass flow rate over 50 s indicates that 7.2 kg have been consumed (Figure Example 2.4.1).

CONSERVATION OF MOMENTUM

For momentum, the basic conservation law for a closed system is of vectorial nature. It relates the momentum vector $\vec{p}$ to the vectorial sum of the forces $\vec{F}$ acting on the system. The change of momentum of the fixed mass equals the resultant of the applied force:

$$\frac{D\vec{p}}{Dt} = \sum \vec{F} \tag{2.8}$$

We take as our intensive property b a vector, namely the velocity vector, and as extensive property B the momentum vector. Hence,

$$B = \vec{p} = m \cdot \vec{V} = m \cdot b$$

Replacing in Equation 2.6 the foregoing definitions, we get

$$\frac{D\vec{p}}{Dt} = \frac{d\vec{p}_\sigma}{dt} + \dot{m}_o \cdot \vec{V}_o - \dot{m}_i \cdot \vec{V}_i$$

Since Equation 2.8 shows that the change of momentum of the closed mass equals the vectorial sum of the forces applied to it, we have, for the CV,

$$\sum \vec{F} = \frac{d\vec{p}_\sigma}{dt} + \dot{m}_o \cdot \vec{V}_o - \dot{m}_i \cdot \vec{V}_i \tag{2.9}$$

or, the sum of the applied forces equals the change in momentum of the mass in the CV, plus the rate of momentum exiting the system, minus the rate of momentum entering the system.

Example 2.5 (VisSim) shows how to use Equation 2.9 to calculate the position and velocity of a rocket ejecting mass as per the schedule of Example 2.4.

EXAMPLE 2.5 ROCKET ACCELERATION, VELOCITY, AND POSITION [VISSIM]

A rocket in outer space (no external forces acting on it) fires its engine with the schedule in Example 2.4. The mass exits the rocket with velocity V_o. Calculate the acceleration, velocity, and distance from the original position of the rocket.

$$V_o = -1000 \text{ m/s (Figure Example 2.5.1)}$$

Solution

Since no external forces are applied to the CV and no mass comes in, we have

$$\sum \vec{F} = \frac{d\vec{p}_\sigma}{dt} + \dot{m}_o \cdot \vec{V}_o - \dot{m}_i \cdot \vec{V}_i \tag{2.9}$$

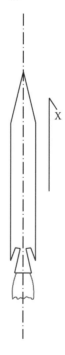

FIGURE EXAMPLE 2.5.1 A rocket moving in 1-D.

We drop the vector notation because all vectors are aligned along x. Then,

$$\frac{dp_\sigma}{dt} = \frac{d(M(t) \cdot V(t))}{dt} = M(t) \cdot \frac{dV(t)}{dt} + V(t) \cdot \frac{dM(t)}{dt} = -\dot{m}_o(t) \cdot V_o$$

We know the derivatives $dM(t)/dt$ (calculated in Example 2.4) and mfr_o, but not the velocity $V(t)$ or acceleration $dV(t)/dt$. However, the acceleration is given from the foregoing equation as

$$\frac{dV(t)}{dt} = \frac{-\dot{m}_o(t) \cdot V_o - V(t) \cdot \dfrac{dM(t)}{dt}}{M(t)}$$

Once the acceleration $dV(t)/dt$ is known, double integration yields the velocity and then the position along x. The block diagram for the RHS of the acceleration is (VisSim, Example 2.4).

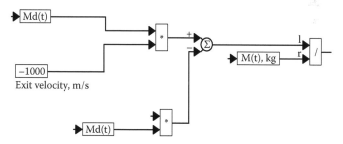

where $Md(t)$ is the $dM(t)/dt$ calculated in Example 2.4, and the mass of the rocket is calculated also from the same example using DELTA, the amount ejected by time t, as follows.

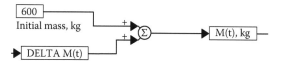

The acceleration is integrated twice to yield velocity and position, which can be readily plotted (Figure Example 2.5.2)

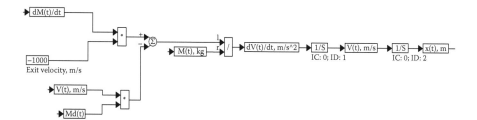

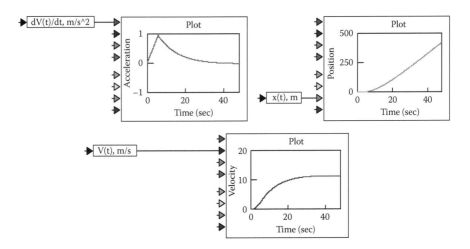

FIGURE EXAMPLE 2.5.2 Acceleration, position, and velocity of the rocket.

In a frictionless world such as the one postulated here, the rocket attains a constant velocity that will stay constant, since no forces oppose the motion. Our daily reality shows this not to be the case, with irreversibilities eventually dissipating kinetic energy.

CONSERVATION OF ENERGY

We derive another form of the First Law, suitable for CVs in transient mode. Our fundamental law for a closed system of mass **sm** is Equation 1.13 (Chapter 1):

$$\frac{dE}{dt} = \frac{dQ}{dt} - \frac{dW}{dt} \qquad (1.13)$$

To apply Equation 2.6 in order to obtain the First Law for an open system, we take

$$B = E = m \cdot e = m \cdot b$$

and application of Equation 2.6 to the system energy yields

$$\frac{DE}{Dt} = \frac{dE_\sigma}{dt} + \dot{m}_o \cdot e_o - \dot{m}_i \cdot e_i$$

The rate of work and of heat transfer of the closed system **sm** equals those of the CV when **sm** is contained in the CV. Then, from the foregoing and Equation 1.14, we have

$$\frac{DE}{Dt} = \frac{dQ}{dt} - \frac{dW}{dt} = \frac{dE_\sigma}{dt} + \dot{m}_o \cdot e_o - \dot{m}_i \cdot e_i \qquad (2.10)$$

The energy per unit mass e is given by Equation 2.1. The power exchanged by the CV merits special consideration. When the unit mass is swept by a pressure p_i into a system, by virtue of Equation 2.3, we know that the product $p_i \cdot v_i$ is the work done per unit mass. Multiplying the work per unit mass times the mass per unit time, we get the work per unit time (i.e., the power) required to ensure that the mass flow rate enters the CV at the given pressure. A minus sign reflects that this work is done on the system. This work is called the expansion work, that is,

$$\frac{dW_{exp,i}}{dt} = -\dot{m}_i \cdot p_i \cdot v_i \tag{2.11}$$

An analogous expression for the exit work is

$$\frac{dW_{exp,o}}{dt} = \dot{m}_o \cdot p_o \cdot v_o \tag{2.12}$$

The rate of work exchanged by the CV can then be cast as the sum of shaft power $\frac{dW_\sigma}{dt}$ (and/or magnetic, electrical, etc.) and expansion work, namely

$$\frac{dW}{dt} = \frac{dW_\sigma}{dt} + \dot{m}_o \cdot (p_o \cdot v_o) - \dot{m}_i \cdot (p_i \cdot v_i) \tag{2.13}$$

Combining the foregoing equations and Equation 2.10, we obtain

$$\frac{dQ}{dt} - \frac{dW_\sigma}{dt} = \frac{dE_\sigma}{dt} + \dot{m}_o \cdot \left(u + p \cdot v + \frac{V^2}{2} + g \cdot z \right)_o - \dot{m}_i \cdot \left(u + p \cdot v + \frac{V^2}{2} + g \cdot z \right)_i$$

which, upon substitution of the definition of enthalpy (Equation 2.3), yields

$$\frac{dQ}{dt} - \frac{dW_\sigma}{dt} = \frac{dE_\sigma}{dt} + \dot{m}_o \cdot \left(h + \frac{V^2}{2} + g \cdot z \right)_o - \dot{m}_i \cdot \left(h + \frac{V^2}{2} + g \cdot z \right)_i \tag{2.14}$$

where the work in Equation 2.14 is any work (shaft work, electrical, or even expansion when not associated with a mass flow rate) that crosses the CV boundaries.

Example 2.6 (VisSim) shows how to use Equation 2.14 to calculate the power and energy delivered under steady state by a CV.

EXAMPLE 2.6 STEADY STATE ENERGY EXTRACTION FROM A STREAM [VisSim]

An adiabatic system delivers power P, constant over a period τ, during which the system remains in steady state. The inlet and exit enthalpies, KE and PE, of the system are also given, as is the constant mass flow rate. Determine

    a. The power output
    b. The total mechanical energy and mass throughput delivered over the period τ

$\dot{m} = 12000$ kg/s
$h_i = 2700$ kJ/kg $h_o = 60$ kJ/kg
$V_i = V_o = 3$ m/s
$z_i = 200$ m $z_o = 2$ m
$\tau = 3890$ s

Solution

We apply Equation 2.14, which, for steady state and no heat exchanged, reads

$$-\frac{dW_\sigma}{dt} = -P = \dot{m} \cdot \left[\left(h + \frac{V^2}{2} + g \cdot z \right)_o - \left(h + \frac{V^2}{2} + g \cdot z \right)_i \right] \tag{a}$$

The enthalpy KE and PE of the stream is easily constructed for the inlet and merged into a compound block (INLET H, KE, PE) for simplicity.

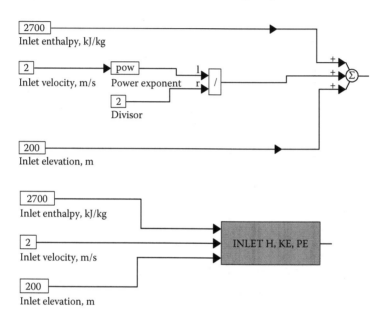

The power output calculation is given by Equation (a) as

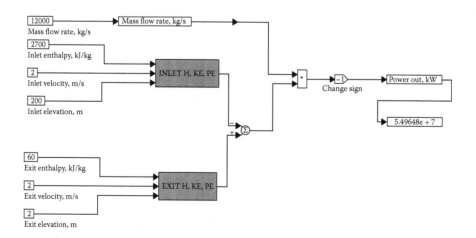

Integration of the power and mass flow rate yields the total energy and mass delivered in the time interval τ.

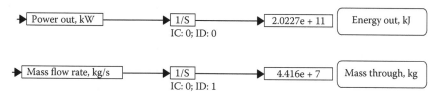

Notice that this example details a hypothetical energy conversion. The inlet stream flows from a state of high enthalpy and potential energy to a state of low enthalpy and potential energy. The energy loss per unit time appears as power output. Hence, the First Law analysis simply reflects an energy conversion from enthalpy, KE and PE, to mechanical or electrical, without the agency of heat production or consumption. Irreversibilities would result in a loss of work and in higher temperatures of the exit stream, but the problem as formulated does not give enough information on how to calculate this.

Example 2.7 (VisSim) shows how to use Equation 2.14 to calculate the power and energy delivered by the CV of Example 2.5 in transient regime.

EXAMPLE 2.7 STREAM MASS FLOW RATE VARYING WITH TIME [VISSIM]

An adiabatic system delivers power P, over a period τ, during which the system remains in steady state. The inlet and exit enthalpies, KE and PE, of the system are also given, as is the mass flow rate, which varies with time. Determine

    a. The power output
    b. The total mechanical energy and mass throughput delivered over the period τ

$\dot{m} = 12000 \cdot \text{kg/s} \cdot (1 + \sin(0.002 \cdot t))$
$h_i = 2700 \text{ kJ/kg} \quad h_o = 60 \text{ kJ/kg}$
$V_i = V_o = 3 \text{ m/s}$
$z_i = 200 \text{ m} \quad z_o = 2 \text{ m}$
$\tau = 3890 \text{ s}$

Solution

We apply Equation 2.14, which, for the CV in steady state, no heat exchanged, and variable mass flow rate, reads

$$-\frac{dW_\sigma}{dt} = +\dot{m}(t) \cdot \left[\left(h + \frac{V^2}{2} + g \cdot z \right)_o - \left(h + \frac{V^2}{2} + g \cdot z \right)_i \right] \tag{a}$$

The only difference between this example and Example 2.6 is the variable flow rate, implemented using the blocks shown,

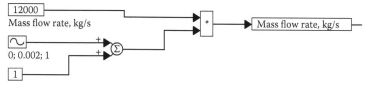

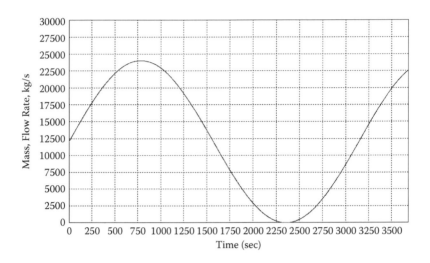

FIGURE EXAMPLE 2.7.1 The mass flow rate varies with time.

which gives mass flow rate temporal profile copied as shown from VisSim (Figure Example 2.7.1).

The power output varies similarly with time, as can be inspected with the viewer.

ENTROPY GENERATION

We can also use Equation 2.6 to track entropy changes. Our basic definition is Equation 1.15 (Chapter 1), namely, for the closed system of mass **sm**:

$$\frac{dS}{dt} = \frac{1}{T} \cdot \frac{dQ_{rev}}{dt}$$

(1.15)

which, in the case of multiple surfaces exchanging heat, can be written as

$$\frac{dS}{dt} = \sum_{k} \frac{1}{T_k} \frac{dQ_{rev}}{dt}$$

Hence, taking the following definitions

$$B = S = sm \cdot b$$

Equation 2.6, for a number of surfaces k bounding the CV and being traversed by heat flow, becomes, when **sm coincides with the CV**,

$$\frac{DS}{Dt} = \sum_{k} \frac{1}{T_k} \frac{dQ_{rev}}{dt} = \frac{dS_{\sigma}}{dt} + \dot{m}_o \cdot s_o - \dot{m}_i \cdot s_i$$

which is written more appropriately as

$$\frac{dS_\sigma}{dt} = +\dot{m}_i \cdot s_i - \dot{m}_o \cdot s_o + \sum_k \frac{1}{T_k} \frac{dQ_{rev}}{dt}$$

(2.15)

Hence, the entropy of the mass in the CV varies as mass carries entropy in or out, and the heat transfer across the boundaries (assumed reversible) increments the entropy when the CV receives heat and reduces it when heat is lost.

Example 2.8 (VisSim) uses Equation 2.15 to show that the entropy of σ does not change for an adiabatic system in periods of steady state even if transient demands exist.

EXAMPLE 2.8 ENTROPY CHANGES WHEN MIXING HOT AND COLD WATER [VISSIM]

A water heater responds to a demand that varies randomly with time. Hot water leaves the heater (called the tank), and it is mixed with cold water to obtain the delivery temperature. We calculate the entropy change of the control volume σ.

The demand can be generated using a sample and hold block (S&H) and a random Gaussian distribution. The program generates random numbers distributed around zero. When those numbers are greater than one, the S&H block changes status. The outcome is a series of pulses of varying duration, which we take to define the demand flow rate.

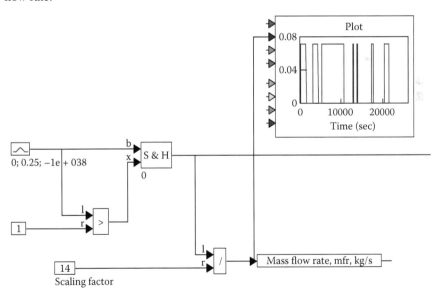

Other data are the supply water temperature (280 K), the delivery temperature (330 K), and the temperature of the water in the tank (380 K).

It is necessary to calculate the entropy variation in the control volume around the mixing chamber where water from the tank with mass flow rate *mfrh* is mixed with supply water of mass flow rate *mfrc* to adjust the delivery temperature of the delivered stream *mfr*.

Solution

Application of Equation 2.14 to the adiabatic mixing process with no work and no kinetic or potential energy gives

$$0 = \frac{dE_\sigma}{dt} + mfr \cdot h - mfrc \cdot h_c - mfrh \cdot h_h$$

We now adopt a rather far-reaching simplification; we assume that the fluid in the mixing chamber adjusts instantly to maintain constant temperature. Then, the energy within σ is constant, and the energy balance reads

$$0 = +mfr \cdot h - mfrc \cdot h_c - mfrh \cdot h_h$$

The mass conservation for the control volume reads

$$0 = +mfr - mfrc - mfrh$$

Combining the last two equations, we get

$$mfrh = mfr \cdot \frac{(h - h_c)}{(h_h - h_c)} \qquad (a)$$

The block

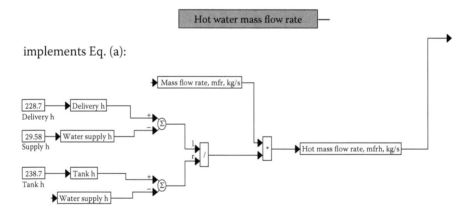

implements Eq. (a):

The cold water flow rate is given by this block:

From Equation 2.15,

$$\frac{dS_\sigma}{dt} = +mfr_i \cdot s_i - mfr_o \cdot s_o + \sum_k \frac{1}{T_k} \frac{dQ_{rev}}{dt} \qquad (2.15)$$

for the adiabatic mixing process, we have

$$\frac{dS_\sigma}{dt} = mfrc \cdot h_c + mfrh \cdot h_h - mfr \cdot h \qquad \text{(b)}$$

Equation (b) is implemented in the following block.

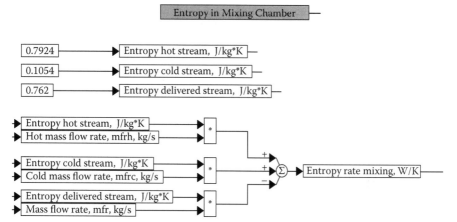

Although anticlimactic, the entropy rate of mixing in this problem is, identically, zero for all times. This should not be surprising, for we have imposed the state of the mixing chamber not to vary with time. Because entropy is a state property, no variations of entropy with time should be expected if the state of σ does not change.

The entropy of σ may or may not be constant, yet there is no conservation in the topic of entropy: like an omnipresent meter, entropy measures thermal change. To fully comprehend this metering function, consider our CV σ exchanging heat with another system $\sigma1$ that furnishes or receives the mass flows, the work and the heat released by σ (Figure 2.4). The entropy generated in σ is

$$\frac{dS_\sigma}{dt} = +\dot{m}_i \cdot s_i - \dot{m}_o \cdot s_o - \frac{1}{T_\sigma}\frac{dQ_\sigma}{dt} \qquad \text{(a)}$$

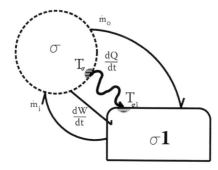

FIGURE 2.4 CVs exchanging mass, heat, and work. The net contribution to entropy increase ensues from heat transfer.

where the negative sign for the heat rate indicates heat loss, and dQ_σ/dt is hence carried as positive. For σ1, we have

$$\frac{dS_{\sigma 1}}{dt} = -\dot{m}_i \cdot s_i + \dot{m}_o \cdot s_o + \frac{1}{T_{\sigma 1}} \frac{dQ_{\sigma 1}}{dt} \qquad (b)$$

Adding (a) and (b) for the total entropy variation gives us

$$\frac{dS_{tot}}{dt} = \left(\frac{1}{T_{\sigma 1}} - \frac{1}{T_\sigma} \right) \frac{dQ}{dt} \qquad (2.16)$$

The Second Law assures us that heat flows from hot to cold temperatures, so the bracket must be positive (dQ/dt is also positive) and

$$\frac{dS_{tot}}{dt} > 0 \qquad (2.17)$$

Equations 2.15 and 2.17 tell us that, no matter what we do or how we do it, entropy generation will be positive for the CV and the surrounding CV for real processes that convert mechanical energy into heat.

Example 2.9 (VisSim) uses Equation 2.16 to show how the total entropy increases due to heat.

EXAMPLE 2.9 ENTROPY GENERATION [VISSIM]

Calculate the entropy generation rate of Example 2.8 if the water heater receives make-up heat from a source at hotter temperature and has a convective heat loss.

Hot source temperature: $T_H = 1500 \text{ K}$

To calculate the convective heat loss, the following is available:

Thermal resistance, water to ambient: $R_{w-amb} = 0.074 \dfrac{\text{m}^2 \cdot \text{K}}{\text{W}}$

Ambient temperature: $T_{amb} = 290 \text{ K}$

Tank area: $At = 1 \text{m}^2$

Solution

When hot water is drawn from the tank, cold water must make up the mass withdrawn. The make-up heat from the hot source is calculated in the block which calculates the heat input required to bring the make-up water to the tank temperature.

$$\boxed{\text{Make-up heat}}$$

Again, we assume this process to happen in such a way that the tank temperature stays constant. The tank is constantly losing heat to the ambient. This standby loss is computed in block using techniques already explored.

$$\boxed{\text{Stand} - \text{by losses}}$$

As shown with reference to Equation 2.16, the rate of entropy generation that it calculates in CVs is exclusively due to heat flow. The heat transfer operation from the hot source to the tank generates entropy. Similarly, heat transfer from the tank to the ambient generates entropy. Another source of entropy generation is internal irreversibilities, explored in the next example. Equation 2.16 is implemented for the two heat transfer operations as follows. The block is

Entropy Generation Rate

The entropy generated by the transfer hot source to tank is

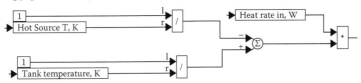

Identically, for the standby loss, we have

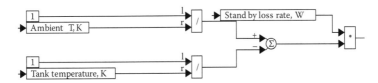

The sum of both entropy generation rates is the overall entropy generation rate in the heater, and it is always positive, as one would expect from Equation 2.16; it is shown in Figure Example 2.9.1.

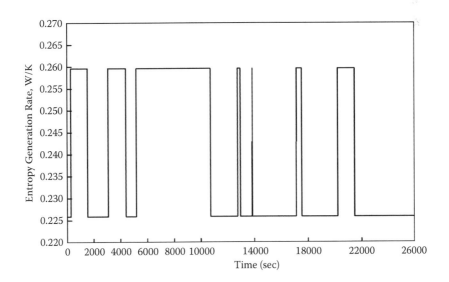

FIGURE EXAMPLE 2.9.1 Entropy generation rate as heater operates.

Example 2.10 uses Equation 2.15 to show how the total entropy increases due to internal irreversibilities in systems with no heat transfer.

EXAMPLE 2.10 ENTROPY GENERATION IN ADIABATIC IRREVERSIBLE PROCESS [VISSIM]

Calculate the total entropy generation rate when air flows through a turbine in steady state. Pertinent data are

Inlet, i $p = 2.0$ MPa $T = 1500$ K $h = 1636.0$ kJ/kg $s = 7.752$ kJ/kg·K

Exit, o $p = 0.1$ MPa $T = 1173$ K $h = 877.4$ kJ/kg $s = 7.952$ kJ/kg·K

$\dot{m} = 1$ kg/s

Solution

For the entropy change associated with the turbine, the following equation applies:

$$\frac{DS}{Dt} = \sum_k \frac{1}{T_k} \frac{dQ_{rev}}{dt} = \frac{dS_\sigma}{dt} + \dot{m}_o \cdot s_o - \dot{m}_i \cdot s_i \tag{a}$$

Because the turbine works in steady state,

$$\frac{dS_\sigma}{dt} = 0$$

which implies from (a) that

$$\frac{DS}{Dt} = \sum_k \frac{1}{T_k} \frac{dQ_{rev}}{dt} = \dot{m}_o \cdot s_o - \dot{m}_i \cdot s_i$$

The internal heating of the fluid is such that

$$s_o > s_i$$

as can be ascertained from the given data. Then,

$$\frac{DS}{Dt} = \sum_k \frac{1}{T_k} \frac{dQ_{rev}}{dt} = \dot{m} \cdot (s_o - s_i) = 1 \cdot \frac{kg}{s} \cdot (7.952 - 7.752) \cdot \frac{kJ}{kg \cdot K} = 0.2 \frac{kW}{K}$$

So, even if no heat transfer occurs, the entropy increase associated with irreversibilities increases the entropy of the air mass, although the entropy of σ stays constant with time. Equation (a) indicates that the entropy increase can be explained by invoking reversible heat transfer into the air flowing inside the turbine. Naturally, the heat input would have to occur at different temperatures, for, as air expands in the turbine producing work, its temperature decreases.

NO "LOST AND FOUND" FOR LOST WORK

Attempts to find a lost object or path can be successful; a great part of the human endeavor goes that way, even if the results have never been, to this point, guaranteed. Yet, entropy increases are unavoidable and irreversible, and they entail a loss of mechanical work each time a transformation occurs. This work cannot be retrieved once the transformation is complete; the opportunity to produce this work is gone.

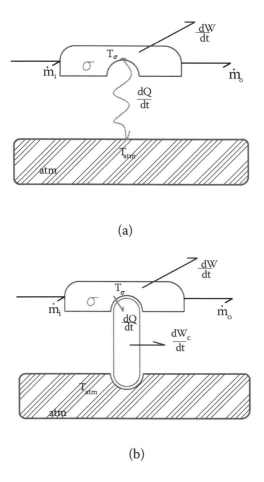

FIGURE 2.5 Heat transfer between a system and the atmosphere.

To see how the lost mechanical power can be related to lost heat, consider the CV σ and the surroundings (atm) (Figure 2.5 (a)), which we assume to be at a constant temperature T_{atm}.

The heat is flowing from T_{σ} to T_{atm}. When the heat is transferred across a temperature difference, the potential to do work exists. Maximum work would be obtained by placing a Carnot engine between the system and the atmosphere (Figure 2.5 (b)). From Equation 1.19 and Figure 2.5, the power produced would be

$$\frac{dW_{Ca}}{dt} = \frac{dQ}{dt} \cdot \left(1 - \frac{T_{atm}}{T_{\sigma}}\right)$$

This expression can be written by virtue of Equation 2.16, for the system *atm* instead of $\sigma 1$, as

$$\frac{dW_{Ca}}{dt} = \frac{dQ}{dt} \cdot T_{atm} \cdot \left(\frac{1}{T_{atm}} - \frac{1}{T_{\sigma}}\right) = T_{atm} \cdot \frac{dS_{tot}}{dt}$$

Adding both the foregoing equations yields

$$\frac{dW_{Ca}}{dt} = \frac{dW_{lost}}{dt} = T_{atm} \cdot \frac{dS_{tot}}{dt} \tag{2.18}$$

When heat transfer occurs, and no attempt is made to recover any of this heat, a penalty is incurred: work is lost, and the rate at which work is lost is proportional to the rate of entropy generation. The optimistic viewpoint here would be that at least the loss has an upper limit. The pessimistic view would be to focus on the fact that entropy always increases and hence power (or the opportunity to extract it) is always being lost. The practical view indicates that energy on the planet is finite, that an intelligent species owes the best of planning to its offspring, and that useless entropy increases deplete energy stores. Consequently, it is useful to obtain a direct estimate of the power lost, which is essentially proportional to the rate of entropy increase.

Example 2.11 (VisSim) shows power and lost power for the engine of Example 1.8 (Chapter 1).

EXAMPLE 2.11 ENTROPY GENERATION AND LOST WORK [VISSIM]

Determine the entropy generation rate and the work lost during one cycle of the engine of Example 1.8 (Chapter 1). Consider that the cycle receives heat from a hot source at 1500 K and rejects heat to the atmosphere at 300 K. The temperature at which heat is accepted by the engine is given by (Ti corresponds to one cycle of this engine)

$$Tsigma(t) = 900 + 205 \cdot \cos\left(\frac{2 \cdot \pi}{Ti}\right) \quad 0 \leq t \leq 0.0158 \text{ s}$$

and the temperature at which heat is rejected is

$$Tsigma(t) = 890 + 205 \cdot \cos\left(\frac{2 \cdot \pi}{Ti}\right) \quad 0.0158 \text{ s} \leq t \leq Ti$$

Solution

The data compound block defines necessary data as shown in the block

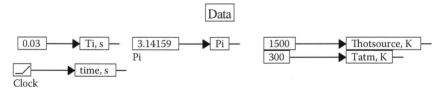

Ti is the cycle duration. The equation for work developed is (Example 1.8)

$$\frac{dW}{dt} = Wdot(t) = \left[33 + 150 \cdot \sin\left(\frac{2 \cdot \pi}{Ti} \cdot t\right) + 35 \cdot \sin\left(\frac{4 \cdot \pi}{Ti} \cdot t\right) + 21 \cdot \sin\left(\frac{5 \cdot \pi}{Ti} \cdot t\right)\right] W$$

This equation is implemented in the following block

$$\boxed{Wdot, W} \!\!-\!\!$$

which contains the following operations.

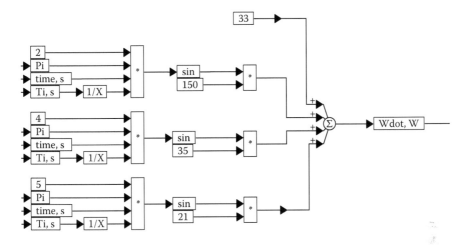

Similarly, the rate of heat exchanged (Example 1.8)

$$\frac{dQ}{dt} = Qdot(t) = \left[35.674 + 205 \cdot \cos\left(\frac{2 \cdot \pi}{Ti} \cdot t \right) \right] W$$

is implemented in the block

Qdot,W

which contains the following operations

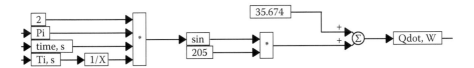

The system temperature is given by the compound block as

Tsigma, K —

which contains the following operations

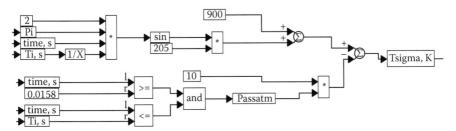

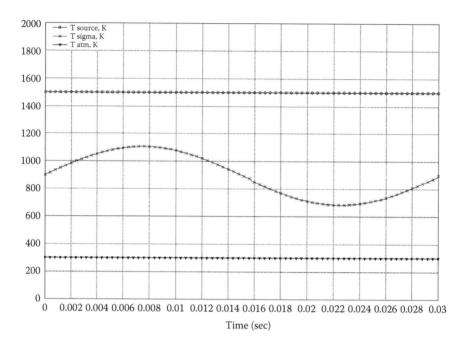

FIGURE EXAMPLE 2.11.1 Source, system, and sink temperatures.

The temperature of the system versus time, compared to the hot and cold source, is shown in Figure Example 2.11.1.

Equation 2.16 allows calculation of the entropy generation rate, done in the block Stotdot as follows:

$$\boxed{\text{Stotdot}} \text{ ---}$$

which contains the following operations

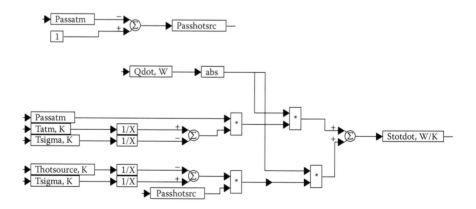

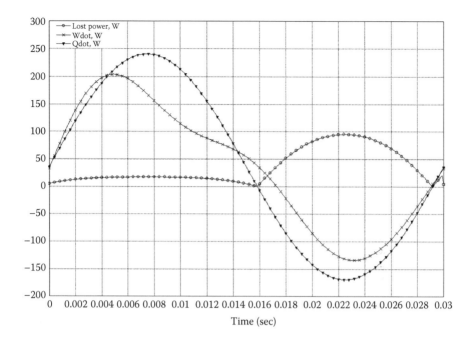

FIGURE EXAMPLE 2.11.2 Power and lost power for the engine.

The variables *Passatm* and *Passhotsrc* are simple filters that equal "1" when heat is transferred to the atmosphere or received from the hot source.

The power lost is calculated via Equation 2.18.

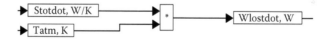

Plotting the power lost and the power output shows that most of the power is lost during the heat rejection cycle (Figure Example 2.11.2).

The integral of lost power over a cycle is nearly equal to the net power delivered

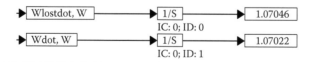

EXERGY, OR WORK NOT YET LOST

Consider a stream going through a reversible contraption (Figure 2.6). The contraption extracts an amount of reversible work $dW_{\sigma,rev}$ from the stream, and exchanges dQ with the atmosphere. We drop all references to time, including the derivatives, since reversible machines work so slowly that time becomes insignificant. We also assume

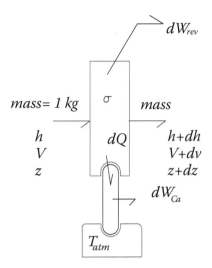

FIGURE 2.6 A reversible device receives 1 kg of mass over a very long time, producing reversible work and activating a Carnot engine.

the mass undergoing the process to be the unit mass, or 1 kg. As per Equation 2.14, the First Law for the unit mass through σ indicates that

$$dQ - dW_{\sigma,rev} = (h+dh) - h + \left(\frac{V^2}{2} + d\frac{V^2}{2}\right) - \frac{V^2}{2} + g\cdot(z+dz) - g\cdot z$$

which readily simplifies to

$$-dW_{\sigma,rev} = dh - dQ + d\frac{V^2}{2} + g\cdot dz \qquad\qquad (a)$$

For the process to be truly reversible, the heat dQ must be used by a Carnot engine so that the engine will add dW_{Ca} to the work $dW_{\sigma,rev}$ produced by σ (Equation 1.19)

$$dW_{Ca} = dQ_e \cdot \left(1 - \frac{T_{atm}}{T_\sigma}\right) \qquad\qquad (b)$$

which can be written as

$$dW_{Ca} = dQ_e - T_{atm} \cdot \frac{dQ_e}{T_\sigma} \qquad\qquad (c)$$

Since the heat received by the engine is heat lost by σ (namely, $dQ_e = -dQ$), we have

$$dW_{Ca} = -dQ + T_{atm} \cdot \frac{dQ}{T_\sigma} \qquad\qquad (d)$$

Using Equation 2.15 for the definition of entropy, we obtain (e):

$$-dW_{Ca} = dQ - T_{atm} \cdot dS_\sigma - T_{atm} \cdot ds \qquad\qquad (e)$$

which, added to Equation (a), results in

$$-dW_{\sigma,rev} - dW_{Ca} = dh + d\frac{V^2}{2} + g\cdot dz - T_{atm} \cdot (dS_\sigma + ds) \qquad\qquad (f)$$

The negative of the total reversible work obtainable from the unit mass is called exergy (*ex*) [2], and hence the change in exergy in this infinitesimal process is

$$d(ex) = dh + d\frac{V^2}{2} + g \cdot dz - T_{atm} \cdot (ds + dS_\sigma) \tag{2.19}$$

When the entropy change of the mass inside σ is null (i.e., when steady state occurs and all the mass in σ stays in the same state, with the same entropy during the process), Equation 2.19 simplifies to

$$d(ex) = dh + d\frac{V^2}{2} + g \cdot dz - T_{atm} \cdot (ds) \tag{2.20}$$

where all the properties belong to the stream, not to the mass inside σ. If the stream were to be brought to rest at the reference z and in equilibrium with the atmosphere, it would lose all its capacity to do work. This is called the dead state of zero exergy. By integration of Equation 2.20 between the dead state and a generic state with enthalpy $h > h_{atm}$, velocity V, entropy s, and elevation z, we find that

$$ex = (h - h_{atm}) + \frac{V^2}{2} + g \cdot z - T_{atm} \cdot (s - s_{atm}) \tag{2.21}$$

Equation 2.20 shows the maximum work to be obtained from the unit mass. Example 2.12 calculates the actual work extracted from, and the exergy of, a stream in an aircraft turbine.

EXAMPLE 2.12 EXERGY OF A STREAM

An airplane is flying high at altitude H. A stream of gas, which we assume to be air, enters the turbine of the gas turbine engine powering the plane. The turbine (Figure Example 2.12.1) extracts power from the stream in steady state, discharging the stream to the atmosphere. Compare the power produced to the power corresponding to the stream exergy.

At the turbine inlet, the stream has the following properties:

$T1 = 1800$ K $h1 = 2003.3$ kJ/kg $p1 = 3$ MPa $s1 = 2.692$ kJ/kg · K

$H = z1 = 5000$ m $V1 = 200$ m/s

$\dot{m} = 1$ kg/s

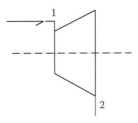

FIGURE EXAMPLE 2.12.1 Turbine schematic.

The discharge of the turbine is at atmospheric pressure and, assuming an ideal, isentropic process, has the following values:

$T2 = 781.4$ K $h2 = 801.6$ kJ/kg $p2 = 0.1$ MPa $s2 = s1 = 2.692$ kJ/kg · K
$H = z2 = 5000$ m $V2 \sim 0$ m/s

The atmosphere is at $T_{atm} = 300$ K

and the air properties are $h_{atm} = 300.2$ kJ/kg $s_{atm} = 1.702$ kJ/kg · K

The turbine efficiency (i.e., the ratio of actual to isentropic power) is $\eta = 0.92$

Solution

The First Law, Equation 2.14, reads

$$\frac{dQ}{dt} - \frac{dW_\sigma}{dt} = \frac{dE_\sigma}{dt} + \dot{m}_o \cdot \left(h + \frac{V^2}{2} + g \cdot z \right)_o - \dot{m}_i \cdot \left(h + \frac{V^2}{2} + g \cdot z \right)_i$$

and applied to our adiabatic turbine in steady state with no elevation change, we have

$$-\frac{dW_\sigma}{dt} = \dot{m} \cdot \left[\left(h + \frac{V^2}{2} \right)_2 - \left(h + \frac{V^2}{2} \right)_1 \right]$$

Since the exit kinetic energy is negligible, we can then write

$$-\frac{dW_\sigma}{dt} = \dot{m} \cdot \left[(h_2) - \left(h_1 + \frac{V_1^2}{2} \right) \right]$$

$$= (1) \cdot \left[801.6 - 2003.3 - \left(\frac{200^2}{2} \right) \right] = -1.203 \text{ MW}$$

The value of 1.203 MW corresponds to an isentropic turbine, namely ($s1 = s2$). Because of internal irreversibilities, mechanical energy is converted into thermal energy. Hence, the flow experiences internal heating, and less power is obtained from the turbine. The actual power is obtained by multiplying the ideal power by the isentropic efficiency, namely

$$\frac{dW_\sigma}{dt}\bigg|_{actual} = 1.203 \cdot \text{MW} \cdot \eta_t = 1.203 \cdot 0.92 = 1.107 \text{ MW}$$

The turbine extracts 1.107 MW of power from the stream. The exhaust stream is at atmospheric pressure, but being hot and at 5000 m elevation, it can still perform work. The work that could be obtained from the stream entering the turbine equals its exergy. The exergy is given by

$$Ex = \dot{m} \cdot \left[(h_1 - h_{atm}) + \frac{V_1^2}{2} + g \cdot z_1 - T_{atm} \cdot (s_1 - s_{atm}) \right]$$

$$= (1) \cdot \left[2003.3 - 300.2 + \frac{200^2}{2} + 9.8 \cdot 5000 - 300 \cdot (2.692 - 1.702) \right]$$

$$= 1.475 \text{ MW}$$

So, whereas the turbine extracts a hefty amount of work, a perfect contraption would manage 33% more work from the stream. The exergy calculations reveal the upper limit of the optimistic expectation for power production.

REVERSIBLE WORK AND REAL WORK

Streams cannot deliver the reversible work because of irreversibilities and technology limitations. Consider starting with

$$dex = -dW_{rev} = dh + d\frac{V^2}{2} + g \cdot dz - T_{atm} \cdot ds \tag{2.20}$$

which, obtaining the infinitesimal work indicated by the negative differential on the left, would require reversible processes. In terms of reversible power delivered in an instant dt, we have

$$\frac{dW_{rev}}{dt} = -\frac{dh}{dt} - \frac{d\frac{V^2}{2}}{dt} - g \cdot \frac{dz}{dt} + T_{atm} \cdot \frac{ds}{dt} \tag{2.21}$$

Assume a real process, with exactly the same enthalpy, elevation, and kinetic energy differentials indicated in Equation 2.20. The stream can exchange heat so as to obtain the same enthalpy variation as in Equation 2.20. Further, assume that the process is such that the stream delivers power. We now can write the actual power delivered when the stream experiences the same infinitesimal changes as

$$\frac{dW}{dt} = -\frac{dh}{dt} - \frac{d\frac{V^2}{2}}{dt} - g \cdot \frac{dz}{dt} + \frac{dQ}{dt} \tag{a}$$

The difference between Equation 2.21 and (a) must be greater than zero because the reversible work will always exceed the actual work. Taking the difference, we obtain

$$\frac{dW_{rev}}{dt} - \frac{dW}{dt} = T_{atm} \cdot \frac{ds}{dt} - \frac{dQ}{dt} > 0 \tag{b}$$

which leads to another interesting result, namely

$$T_{atm} \cdot \frac{ds}{dt} > \frac{dQ}{dt} \tag{2.22}$$

Equation 2.22 indicates that, in an adiabatic process ($dQ/dt = 0$), the stream entropy will always increase. We already noted this characteristic of entropy in Example 2.10. Only for a reversible adiabatic process will the entropy change be null.

DIRECT CONVERSION PROCESSES: IS AN EFFICIENCY CLOSE TO ONE POSSIBLE?

Direct conversion processes are those that, unlike cyclic machines, extract energy from a stream of matter or energy without cyclically converting thermal into mechanical energy. Solar cells, windmills, fuel cells, turbines, electric motors, and a myriad of other components of energy systems are direct energy conversion devices. A uniform definition of efficiency is presently lacking, and hence comparisons of

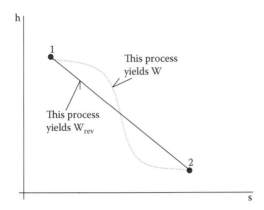

FIGURE 2.7 A stream can follow infinite paths between points 1 and 2. We consider an ideal, reversible one (solid line) and a real one (dotted line).

efficiencies of direct conversion devices are speculative at best. What is clear, though, is that no real device will utilize all the exergy of the input stream, ever, unless the laws that rule the physics of the Universe change.

Consider an adiabatic direct conversion process. Then, by Equation 2.22, the entropy of the stream (or streams) will have to increase to comply with the Second Law. This is important because it leads to the conclusion that only reversible adiabatic direct energy conversion processes will have efficiencies equal to one. All real devices will have efficiencies less than one, no matter how clever or expensive. To see this, consider a process that takes the stream from 1 to 2 (Figure 2.7). The integrated form of Equation 2.21 is, for the reversible process,

$$W_{rev} = (h_1 - h_2)_{rev} + \left[\frac{V1^2 - V2^2}{2} + g \cdot (z_1 - z_2)\right]_{rev} + T_{atm} \cdot (s_2 - s_1)_{rev} \qquad (2.23)$$

Neglecting kinetic and potential energy changes, we have

$$W_{rev} = (h_1 - h_2)_{rev} + T_{atm} \cdot (s_2 - s_1)_{rev} \qquad (a)$$

whereas the actual work is

$$W = h_1 - h_2 \qquad (b)$$

Integration of Equation 2.22 gives, for an adiabatic process (the equality holds for a reversible one),

$$(s_2 - s_1) >= 0 \qquad (c)$$

Adopting as a general definition of efficiency the ratio of actual work to reversible work, and using (a), we have

$$\eta = \frac{W}{W_{rev}} = \frac{(h_1 - h_2)}{((h_1 - h_2)_{rev} + T_{atm} \cdot (s_2 - s_1)_{rev})} \qquad (d)$$

For an adiabatic process, the equality of (c) establishes that (d) is equal to

$$\eta = \frac{W}{W_{rev}} = \frac{(h_1 - h_2)}{((h_1 - h_2) + T_{atm} \cdot (s_2 - s_1)_{rev})} \tag{2.24}$$

Invariably, the efficiency of Equation 2.24 will always be less than one. If the process is not adiabatic, and if $s_2 < s_1$, an efficiency greater than 1, as calculated by the RHS of (d), could result. This would be an enticing goal, but has so far been impossible to achieve. If $s_2 > s_1$, the efficiency will be less than one, as no doubt the inquiring will find in practice.

OTHER USEFUL EMPIRICAL DISSIPATION LAWS

HEAT TRANSFER

Rate laws are, like the First and Second laws, empirical, and have served engineering design for many years. Unlike the First and Second laws, the rate laws require detailed empirical content and sophisticated math for multidimensional solutions. Calculations based on them invariably result in overall entropy increases when properly applied.

As already discussed, heat requires a temperature difference in order to flow. For a temperature distribution ($T(x)$) (Figure 2.8), inside a solid, a tangent line to the distribution exists at every point P. The slope of the tangent line is

$$\frac{dT(x)}{dx} = \tan(\alpha)$$

As long as the slope is different from zero, heat will flow, for a positive or negative slope implies a temperature difference between two points in the solid. The heat flux (heat transferred by conduction per unit area and time), is given in one dimension as proportional to the temperature gradient.

$$\frac{dQ/dt}{dA} = -k \cdot \frac{dT}{dx}$$

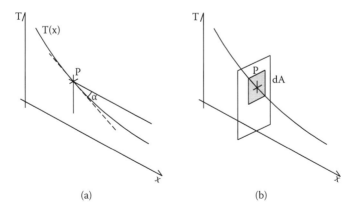

(a)   (b)

FIGURE 2.8 (a) Temperature distribution inside a solid, $T(x)$, and (b) area normal to heat flow.

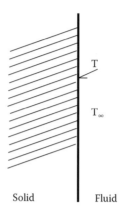

FIGURE 2.9 Heat transfer from a solid to a fluid.

Note that heat flow is positive if the temperature gradient is negative, that is, heat flows from hot to cold regions. The foregoing equation is called Fourier's law. For a solid of constant properties and thickness Δx, the heat flow between two isothermal surfaces 1 and 2 of area A is

$$\frac{dQ}{dt} = k \cdot A \cdot \frac{(T_1 - T_2)}{\Delta x} \tag{2.25}$$

If the heat transfer occurs between a solid surface and a fluid (Figure 2.9), Newton's law of cooling (not of motion) applies:

$$\frac{dQ/dt}{dA} = hc \cdot (T - T_\infty)$$

The heat flows from hot to cold, and hc is the convective heat transfer coefficient, which depends on the properties and flow parameters of the fluid if T is the solid surface temperature. For a finite area A with constant hc, the heat transfer rate is

$$\frac{dQ}{dt} = hc \cdot A \cdot (T - T_\infty) \tag{2.26}$$

When two surfaces at temperatures T_1 and T_2 (Figure 2.10) exchange radiation with each other, energy is transported by electromagnetic waves. Both surfaces emit radiation, and both absorb part of the incoming radiation. If the temperature of the surfaces is different, then net energy transport from the hotter to the cooler surface occurs. The net heat transport is given by

$$\frac{dQ_{12}}{dt} = \sigma \cdot \varepsilon_1 \cdot F_{A1-2} \cdot A1 \cdot \left(T_1^4 - T_2^4\right) \tag{2.27}$$

where σ is the Stefan–Boltzmann constant, and ε_1 is the emissivity of surface 1. The term F_{A1-2} is the view factor from surface 1 to surface 2. The net exchange is proportional to the difference between the fourth power of the surface temperatures, and the surfaces must have constant properties regardless of the radiation

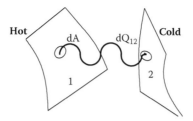

FIGURE 2.10 Two surfaces exchanging radiation.

direction or its wavelength. This independence of direction and wavelength characterizes the so-called gray surfaces. Because any net heat transfer rate requires a temperature difference to occur, and large heat transfer rates are needed in many aspects of technology, it follows that work will be lost when heat transfer takes place.

FRICTION

Another source of irreversibility is friction. Friction manifests itself as a force opposing motion. Between two solid surfaces, the friction force is

$$F_f = \mu \cdot N \tag{2.28}$$

where μ is the friction coefficient between the surfaces, and N is the normal force acting on the surfaces. When an object such as an aircraft moves in a fluid, the drag force due to the combined friction and pressure forces acting on the surface is

$$D = C_D \cdot \frac{\rho \cdot V^2}{2} \cdot A \tag{2.29}$$

where C_D is the drag coefficient, ρ is the fluid density, V is the relative velocity, and A is the area of the body. The drag coefficient depends on the flow characteristics. In the case of flow inside a duct, a net drag force exists opposing the fluid motion. In the case of ducts, it is easier to calculate pressure drops rather than forces. For an internal velocity V, the pressure drop is

$$\Delta p = f \cdot \frac{\rho \cdot V^2}{2} \cdot \frac{L}{Dh} \tag{2.30}$$

where the friction factor f depends on the nature of the flow and the fluid properties, L is the length, and Dh is the diameter or hydraulic diameter of the duct.

RESISTANCE AND ELECTRICAL ENERGY STORAGE

In electrical circuits, irreversibilities manifest themselves as voltage drops. For a resistor R the current is

$$I = \frac{Vr}{R} \tag{2.31}$$

where Vr is the voltage drop to sustain the current I flowing in the resistor R. In a resistor, electrical energy is dissipated at a rate

$$\frac{dex}{dt} = I^2 \cdot R \tag{2.32}$$

The voltage across the terminals of a capacitor is a function of the integral of the current flowing through it. For instance, for a process starting at time 0 and extending into time τ, the voltage is

$$VC(\tau) = \frac{\int_0^\tau I \cdot dt}{C} \tag{2.33}$$

An ideal capacitor does not destroy electrical energy; it simply stores it and returns it quality untouched (i.e., no conversion into heat here). The rate of energy accumulation is

$$\frac{dW}{dt} = \frac{I \cdot \int_0^\tau I \cdot dt}{C} \tag{2.34}$$

The energy is stored by building an electric field between capacitor plates. In an inductor of inductance L, the voltage across terminals is given by

$$VL(t) = L \cdot \frac{dI}{dt} \tag{2.35}$$

An ideal inductor does not dissipate energy either; it simply stores it and returns it. An inductor stores energy by building magnetic fields, typically in the core of coils. The rate at which energy is stored or recovered from an inductance is

$$\frac{dW}{dt} = L \cdot I \cdot \frac{dI}{dt} \tag{2.36}$$

Example 2.13 (VisSim) shows how a resistor generates entropy, whereas ideal capacitors and inductances do not.

EXAMPLE 2.13 A SERIES CIRCUIT AND ENERGY STORAGE [VISSIM]

A circuit has a resistor, capacitor, and inductor in series (RLC) (Figure Example 2.13.1). The applied voltage $V(t)$ is a sinusoid with a frequency of 15 rad/s and amplitude of 1 V.

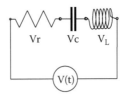

FIGURE EXAMPLE 2.13.1 An RLC circuit.

Other values are

Resistor: 1 Ohm
Capacitor: 0.1 F
Inductance: 0.1 He

1. Determine the current versus time. Also determine the energy stored or dissipated in each component versus time.
2. Assuming the following temperatures, and that the heat dissipated flows freely to the atmosphere, determine the net entropy generation rate

$$\mathrm{Tr} = 1600\,\mathrm{K} \qquad \mathrm{T}_{atm} = 300\,\mathrm{K}$$

Solution

The block "definitions" defines the voltage versus time, and the resistor, capacitor, and inductance values

Definitions

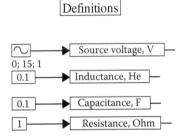

At any instant, the voltage imposed on the series circuit must be equal to the sum of the voltage across each element, namely

$$Vr + VC + VL = V(t)$$

Replacing the above terms with Equations 2.31, 2.33, and 2.35, we have

$$R \cdot i + \frac{\int i \cdot dt}{C} + L \cdot \frac{di}{dt} = V(t)$$

Isolating the first derivative of the current, we find that

$$\frac{di}{dt} = \frac{V(t)}{L} - \frac{R \cdot i}{L} - \frac{\int i \cdot dt}{C \cdot L}$$

This ODE equation can be readily solved by the following block. The first derivative is constructed by adding the resistive and capacitive terms to the first term in the RHS. Simple integration yields the value of the current, which is fed back to calculate the first derivative.

Current, A

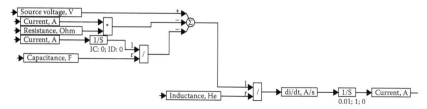

Once the current is known, the power dissipated or being stored is easily calculated from Equations 2.32, 2.34, and 2.36 as follows:

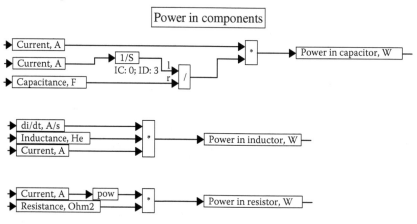

Plotting the power in the inductor and capacitor determined earlier versus time, we obtain Figure Example 2.13.2.

Thus, the net power stored is zero, but power flows out of the inductor into the capacitor and vice versa. Negative power merely entails the inversion of a field (magnetic in the inductor, electric in the capacitor). The resistor simply dissipates positive power. The sum of the stored power plus the dissipated power in the resistor equals the power delivered by the source. The power dissipated by the resistor at Tr flows to the atmosphere as heat. The resistor is assumed to be at constant temperature, and hence, its entropy does not change. The rate of entropy generation is due to heat flow, namely, from Equation 2.15 we have

$$\frac{dS_\sigma}{dt} = +\sum_k \frac{1}{T_k} \frac{dQ_{rev}}{dt} = Power \cdot \left(\frac{1}{T_{atm}} - \frac{1}{Tr} \right)$$

This equation is implemented in the block

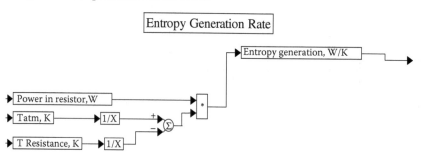

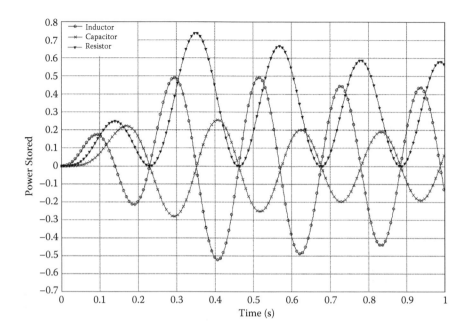

FIGURE EXAMPLE 2.13.2 Power stored or dissipated in the resistor (solid triangle).

The resistor always generates entropy, (i.e., the electrical energy is lost into the low atmospheric temperature) as shown in Figure Example 2.13.3.

Example 2.14 (VisSim) shows how a circuit will respond to an initially charged capacitor with oscillations. The (remote) possibility of storing energy via resonant behavior is touched upon.

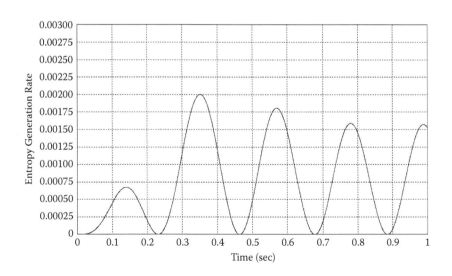

FIGURE EXAMPLE 2.13.3 Resistor entropy generation rate.

EXAMPLE 2.14 RESONANT NETWORKS AND ONE ILLUSION [VISSIM]

The circuit of Example 2.13 is now connected to a DC source (Figure Example 2.14.1). The capacitor gets charged. When the charging process is complete, the battery is disconnected. The switch below the circuit is now closed, and current flows until the capacitor is discharged.

For the values

Resistor: 0.1 Ohm
Capacitor: 0.1 F
Inductance: 0.1 He

and an initial charge in the capacitor of 1 Coulomb, show:

1. Current versus time after the capacitor switch is closed.
2. Energy dissipated/stored versus time after the capacitor switch is closed.

Solution

The current in the circuit at any instant, from Example 2.13, is

$$\frac{di}{dt} = \frac{V(t)}{L} - \frac{R \cdot i}{L} - \frac{\int i \cdot dt}{C \cdot L} \tag{a}$$

The current is related to the charge in the capacitor by

$$i = \frac{dQ}{dt} \tag{b}$$

Replacement of (b) in (a) with $V(t) = 0$ gives

$$\frac{d^2Q}{dt^2} = \frac{V(t)}{L} - \frac{R}{L} \cdot \frac{dQ}{dt} - \frac{Q}{C \cdot L} = -\frac{R}{L} \cdot \frac{dQ}{dt} - \frac{Q}{C \cdot L} \tag{c}$$

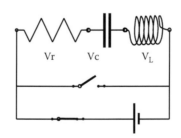

FIGURE EXAMPLE 2.14.1 RLC circuit, with a DC battery.

This equation is now solved subject to the given boundary condition, Q(0) = 1 Coulomb. Values are defined in the block

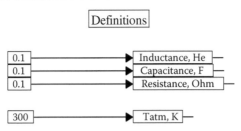

Equation (c) is now solved by assuming that Q and dQ/dt are known to calculate d^2Q/dt^2, after which a double integration (the one yielding Q with the initial condition of 1 Coulomb) yields Q. This is done in the block

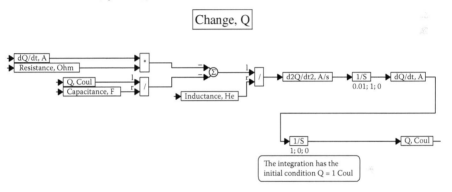

The current (for the R, L, and C values given) follows a decaying oscillatory pattern, typical for the underdamped range of ODEs such as (c). The plot of Figure Example 2.14.2 shows dQ/dt, the current versus time.

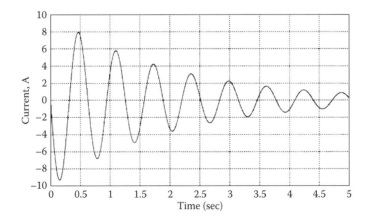

FIGURE EXAMPLE 2.14.2 Current decays with time, as power is dissipated in the resistor.

The power dissipated in each component is easily calculated from Equations 2.32, 2.34, and 2.36 as follows:

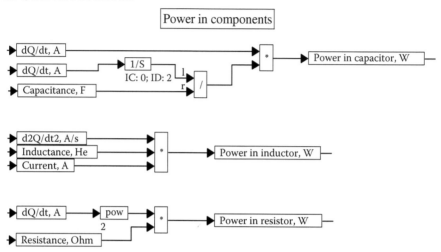

The capacitor and inductor store power (they exchange progressively smaller amounts of power in resonance), whereas the resistor simply dissipates power, decreasing the instantaneous power stored (Figure Example 12.14.3).

The energy dissipated by the resistor must equal the energy originally present in the capacitor, and it does, as shown in the VisSim program. The system oscillates at its natural frequency.

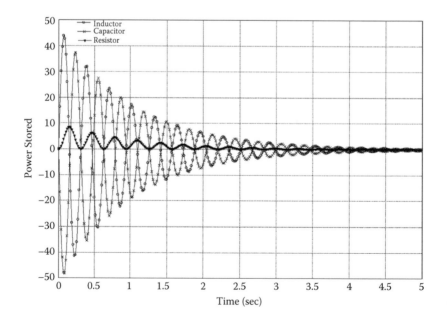

FIGURE EXAMPLE 2.14.3 Power is dissipated in the resistor as oscillations decay.

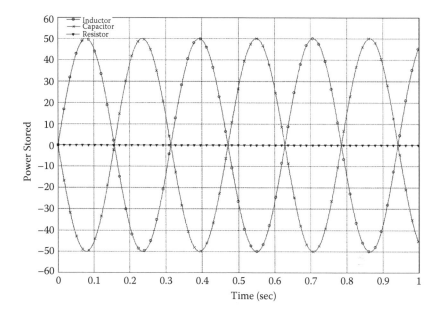

FIGURE EXAMPLE 2.14.4 When the resistance is zero, the circuit resonates and no energy is dissipated.

The resonant behavior can present challenges and opportunities. If a voltage source of the natural frequency of the system were connected to the circuit, the circuit would absorb more and more energy until it limited the source or resulted in destruction of components. In the absence of resistance, the circuit could be a way of storing energy (zero resistance is an impossibility, so a "perfect" energy storage method based on this scheme is a long way off). However, the interesting aspect is that of energy exchange between the capacitor and the inductor, as reflected in the plot when $R = 0$ (Figure Example 2.14.4).

SIMPLE ANALOGIES TO FORMULATE THERMAL PROBLEMS

For practical purposes, Fourier and Newton's laws are often cast in terms of an electrical analogy:

Heat flow is the analog of current.
Temperature difference is the analog of voltage difference.
The analog of $\Delta x / k \cdot A$ (conduction) or $1/h \cdot A$ (convection) is resistance.

Equation 2.31 (Ohm's law) is then applied to the thermal problems for conduction resistance R_{th}:

$$\frac{dQ}{dt} = \frac{\Delta T}{R_{th}} = \frac{(T_1 - T_2)}{\Delta x / k \cdot A} \tag{2.37}$$

or for convection

$$\frac{dQ}{dt} = \frac{\Delta T}{R_{th}} = \frac{\left(T - T_{\infty}\right)}{1/hc \cdot A} \tag{2.38}$$

Fluid flow problems can also be formulated on the basis of Ohm's law. The analogs are

Fluid flow is the analog of current.
Pressure difference is the analog of voltage difference.
A factor proportional to tube length and friction factor is the nonlinear analog of resistance.

From Equation 2.30, it is possible to derive a relationship between pressure drop and mass flow as follows. Multiplying and dividing the RHS of Equation 2.30 by A, the duct area, we have

$$\Delta p = f \cdot \frac{\rho \cdot V^2}{2} \cdot \frac{L}{D} = f \cdot (\rho \cdot V \cdot A) \cdot \left(\frac{V \cdot L}{2 \cdot D \cdot A}\right)$$

which can be written as

$$(\rho \cdot V \cdot A) = \dot{m} = \frac{\Delta p}{\left(\dfrac{f \cdot V \cdot L}{2 \cdot D \cdot A}\right)} = \frac{\Delta p}{R_{flow}(V)} \tag{2.39}$$

Or, the mass flow rate is proportional to the pressure difference available, and inversely proportional to a resistance that depends on the friction factor, the length, and the velocity of the flow. This makes this law highly nonlinear, especially considering that the friction factor is a function of the velocity (or rather the Re number) as well. Other analogies exist in fluid motion [1]. The second-order ODE behavior outlined in Examples 2.13 to 2.14 can be (within limits) found in fluid systems. The topic is taken up again in Chapter 9.

SUMMARY

As streams flow through CVs, their properties change, that is, they change states. The rate of variation of properties can be related to the rate of flow of properties crossing the surface of the CV, as follows:

Mass: $\dfrac{dM(t)}{dt} = \dot{m}_i - \dot{m}_o$ $\tag{2.7a}$

Energy: $\dfrac{dQ}{dt} - \dfrac{dW_\sigma}{dt} = \dfrac{dE_\sigma}{dt} + \dot{m}_o \cdot \left(h + \dfrac{V^2}{2} + g \cdot z\right)_o - \dot{m}_i \cdot \left(h + \dfrac{V^2}{2} + g \cdot z\right)_i$ $\tag{2.14}$

Entropy: $\dfrac{dS_\sigma}{dt} = +\dot{m}_i \cdot s_i - \dot{m}_o \cdot s_o + \sum_k \dfrac{1}{T_k} \dfrac{dQ_{rev}}{dt}$ $\tag{2.15}$

Entropy generation of CV plus surroundings is always positive, which means that work is being lost at a rate proportional to the entropy generation rate. In a broad sense, lost work can be equated to additional fuel expenditure with no useful end.

$$\frac{dW_{lost}}{dt} = T_{atm} \cdot \frac{dS_{gen}}{dt}$$
(2.18)

The ability to do work of a stream of unit mass is given by

$$ex = (h - h_{atm}) + \frac{V^2}{2} + g \cdot z - T_{atm} \cdot (s - s_{atm})$$
(2.21)

If the property values of the stream (h, V, z, or s) come closer to the atmosphere values, its ability to do all types of work (mechanical, electrical, etc., or thermal to mechanical) disappears.

REFERENCES

1. Lewis J. W. 1994. *Modeling Engineering Systems*, High. ISBN 1-878707-08-6.
2. Jones, J. B. and R. Dugan. 1996. *Engineering Thermodynamics*. Prentice Hall, NJ.

3 Predicting Peaks
A Difficult Art

> … long term energy forecasts … have … a manifest record of failure
>
> **V. Smil,** *Energy at the Crossroads*

How much energy (renewable or not) is out there begets (at least) two quali-fications: at what price and with what technology. Obviously, only those equipped to predict the future can intelligently discern these answers. Not being thusly endowed, we address here proven reserves and assess how long they could last at different consumption growth rates. The magnitude of renewable resources is assessed. Speculation about future discoveries and technical advances is enabled so as to delineate what could (and what could not) be effective for stretching supplies. Barring meteorites or the space sta-tion, the mass of the planet remains constant. Not so with the energy released from fossil/nuclear fuels, which, once released, flows to lower temperatures until it is rejected to space. To facilitate appraisal and discussion, we adopt a set of common units, namely metric tonnes of oil equivalent (toe) for energy and year for time. Simulation programs are included on the CD.

DISCLAIMER AND METHOD

We do not attempt here to predict how long any fuel will last. The future is unknown to energy forecasters, and there is plenty of evidence to support this observation [1]. Our objective is much more modest: to show how much energy is reported to be available by credible sources and how long it could last given today's consump-tion patterns. Hopefully, engineers and society will respond as in the past, and as a resource is depleted and its price increases, other sources will be developed. The only aspect that could be different this time is the concern with avoiding environ-mental damage. We also speculate as to the impact of reasonable discoveries or advanced technology on the lifetime of the resource. We cannot ascertain, with any technique, what the future will hold. Arguably, however, we can arrive at conclusions that may serve to guide our work to meet the challenges to come.

It is useful to introduce the widely used parameter called the lifetime of a resource. The *lifetime Li* is defined as the amount of a given resource available (or known to be available), R, divided by its present rate of use *dU/dt*:

$$Li = \frac{R}{\dfrac{dU}{dt}}$$

(3.1)

So, for $R = 10$ units, and $dU/dt = 1$ unit/yr, we have a Li of 10 years. Fossil fuel resources are often evaluated this way. Of course, R is bound to increase in time as new resources are discovered, up to a limit. The limit is reached when the resource is exhausted or when a substitute resource becomes practical. It should be noted that, as the end nears, it is very likely that production rates will decrease as the resource becomes expensive, and one could conclude that Li is indeed healthy when it is not.

Recognizing the limitations of the lifetime concept, a logistical growth model is often invoked. Its basics are explained here. Consider a system initially containing all the available mass T_o of a fuel. If a certain amount is drawn at a rate $\dot{m}_o$, and new resources are found at a rate $\dot{m}_i$, then, from Equation 2.7a (Chapter 2) we find that the system mass $M(t)$ will vary with time from its initial T_o value as per

$$\frac{dM(t)}{dt} = \dot{m}_i - \dot{m}_o$$

Although we will later allow for positive values of $\dot{m}_i$, let us assume for now that no new large discoveries ensue, that is, T_o is essentially constant. Labeling the exit mass flow rate as the rate of use of the resource, we now have

$$\frac{dM(t)}{dt} = -\frac{dU}{dt} \tag{3.2}$$

So, the obvious conclusion is that $M(t)$ decreases as the fuel is used. What is not so obvious is the functional dependence of the production rate dU/dt on time. Awareness of such functional dependence would enable integration of Equation 3.2 and hence a reliable prediction of the lifetime of the fuel.

Logistical growth makes assumptions regarding the dependence of dU/dt versus time. We will see that those assumptions lead to a production peak, regarded as significant by some and as irrelevant by others. At a certain point in time (Figure 3.1), a resource has an extraction rate $P = dU/dt$, namely the average production rate over a year. This rate is small at first but increases as the acceptance of the resource grows.

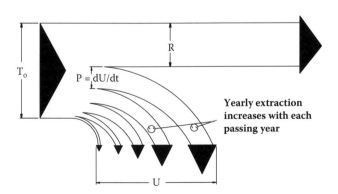

FIGURE 3.1 A model of consumption of a finite resource.

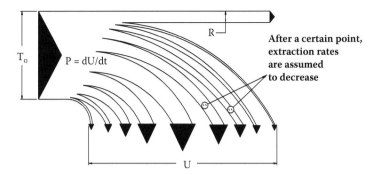

FIGURE 3.2 As less of the finite resource is left, consumption rate R decreases.

The crucial assumption here is that, after half of the resource is gone, the extraction rate P decreases, as in Figure 3.2.

At a later point in time, shown by Figure 3.2 as compared to Figure 3.1, the total amount used, U, has increased (the resource has been used), but R (the amount available) has decreased. Eventually, R may decrease so much that what is left may not be worth recovering for sale because, as the production (and consumption) rate decreases, other sources or alternatives are bound to develop. That other alternatives develop should be clear from history.

The chief assumption involved in the process described by Figure 3.2 is that if production cannot keep up with demand after the midpoint is reached, adjustments (spontaneous or mandated, depending on alternatives) will have to take place. So, an estimation of the midpoint appears to be a useful piece of knowledge for planning purposes.

Let us construct the model mathematically to see what we can ascertain from it. If the extraction rate ($Pr = dU/dt$) is proportional to the amount produced up to the present time U and to the difference between the total amount T_o and U, then

$$\text{Pr} = \frac{dU}{dt} = r \cdot U \cdot (T_o - U) \tag{3.3}$$

The constant of proportionality r is commonly known as the logistic growth rate. In addition to values for r and T_o, this ODE calls for one boundary condition, namely, at some time we must specify the value of U. That is, we must know how much of a resource has been consumed at a certain point in time, or,

$$U = U_o \quad \text{when} \quad t = t_o$$

Note that the reserves R (proven or ultimate, depending on the optimism of the user about the fuel potential) are related to T_o and U as follows:

$$T_o = R + U$$

The simplicity of this approach masks the fact that both T_o and r are bound to be functions of time, as new discoveries (or revisions thereof) and economic conditions change. Because Equation 3.3 can be easily integrated numerically with r and T_o as functions of time to yield some very rough estimates of possible lifetimes, onward we proceed.

Example 3.1 studies Equation 3.3, showing that indeed it reflects the physical situation depicted in Figures 3.1 and 3.2.

EXAMPLE 3.1 LOGISTIC GROWTH

We solve here the logistic growth equation numerically. A closed-form solution exists, but the flexibility of numerical solutions enables variations of parameters versus time that the closed form cannot accommodate. Hence, and to enable creativity in more complex cases, we present a numerical solution to show how a resource can be developed to exhaustion.

Consider a resource with the following characteristics:

Total amount in the ground:	$T_0 = 1000$ tonne
Total extracted in first year:	$U(0) = 1$ tonne
Logistical growth rate:	$r = 0.001$

Equation 3.3 reads:
$$\frac{dU(t)}{dt} - r \cdot U(t) \cdot (To - U(t)) = 0$$

Taking derivatives versus time yields (for constant T_o and r)

$$\frac{d^2U(t)}{dt^2} - r \cdot (To - 2 \cdot U(t)) \cdot \left(\frac{dU(t)}{dt}\right) = 0$$

When the second derivative equals zero, the production can exhibit a peak. This peak occurs then for

$$U = \frac{To}{2}$$

This prediction from analysis, that production peaks when half of the resource is exhausted, is easily verified when Equation 3.3 is solved numerically in MathCad©. We plot here (Figure Example 3.1.1) the production rate (Equation 3.3) (dash) and its temporal integral (solid line).

$$\int_0^\tau \frac{dU(t)}{dt}\, dt = \int_0^\tau r \cdot U(t) \cdot (To - U(t))\, dt = U(\tau)$$

The dashed line shows that the production rate (dU/dt) peaks at 7.5 years, and that by 15 years the cumulative production U (solid line) has reached the T_o value, that is, there is no more to extract. The peak (i.e., the midpoint) is a subject of current debate about oil.

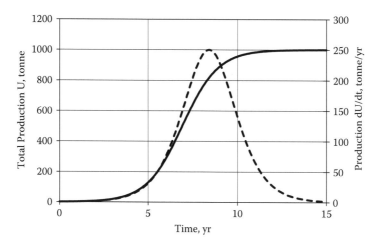

FIGURE EXAMPLE 3.1.1 Production (solid line) and production rate (dashed line).

In this model, two circumstances can change to extend the resource. The growth rate may decrease (reflecting conservation measures leading to more efficient use, or to lower consumption), or the value of T_o may increase due to additional discoveries or enhanced extraction technology. Of course, assessing either implies knowledge of the future, a tricky business. Let us postulate in this example that the growth rate is halved. Then, we obtain the following graph (Figure Example 3.1.2).

This example shows that a sure way to extend the depletion time of a resource is to conserve it. Of course, conserving via enhanced technology is much more fun than conserving via doing without.

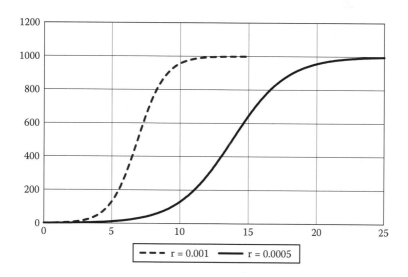

FIGURE EXAMPLE 3.1.2 Production for lower growth rates extends a resource.

Care and good judgment must be adopted when using this (or any) model. Consider for instance some metal industries: minerals get extracted from the Earth, and metals are extracted from minerals (Chapter 6). The metals can be stable (Al, Cu) or can oxidize rapidly (Fe). Whereas the mineral in the ground may be described by the model, metals, once separated, really do not deplete, for their mass on Earth is constant.

Unlike metals, energy resources do deplete: once extracted and oxidized or split, the Second Law (Chapter 2) clearly dictates that reduction or recovery of the original species will require energy.

Equation 3.3 shows that the production rate is zero when the total amount used up is zero ($U = 0$) or when the resource has been used up ($To = U$). The first situation corresponds to time zero, and the second one to the time elapsed to exhaust the resource. In between those instants, the production peaks, as shown in Example 3.1. The peak occurs when half of the resource is exhausted, namely, when

$$U_{peak} = \frac{T_o}{2}$$

The corresponding peak production rate is given by Equation 3.3:

$$\left. \frac{dU}{dt} \right]_{peak} = r \cdot \frac{T_o^2}{2}$$

The conclusions of logistical growth apply only to the systems that follow this pattern. Energy supplies might or might not do this, but the consistent application of this model yields an estimate of resource lifetime that accounts for more meaningful parameters (fuel available, growth rates) than the simple lifetime of Equation 3.1. Numerical techniques used for integration allow meaningful (if speculative) discussions of the effects of reserve growth and speed of growth on resource lifetime.

RESOURCE LIFETIME AND THE LAWS OF THERMODYNAMICS

We have seen that the First and Second Laws describe, respectively, conservation of energy and irreversible energy transformations. Equation 3.3 proposes a way (of many possible ones) whereby a certain amount of mass of fuel in the ground will be consumed, as time flows by. The mass in the ground contains energy in the form of chemical bonds (that are easily oxidized) or nuclei binding energy, hence a direct translation between mass in the ground and energy in the ground is, in principle, possible. The First Law is observed by Equation 3.3 in that, when the value of U reaches the T_o value, the energy in the ground is exhausted. The energy originally in the ground is now somewhere else, for energy is conserved. The Second Law provides an answer as to the ultimate fate of the extracted energy. The process implies

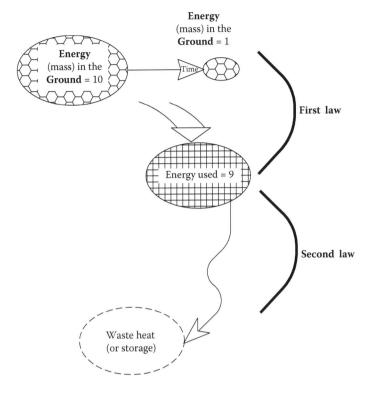

FIGURE 3.3 The ultimate fate of energy extracted from the ground is waste heat.

that, once released, the energy in the ground will flow to less useful energy until its final dissipation into space. The relationship among these concepts is summarized in Figure 3.3.

Figure 3.3 summarizes the First and Second Laws as reflected by Equation 3.3. The sum of the energy used and the energy remaining in the ground at any time will equal the original amount present in the ground, as established by the First Law. As established by the Second Law, these transformations are irreversible, and the ultimate fate of all released energy is waste heat, even though some storage, in the form of potential energy, or bonding energy, is possible.

EXAMPLE OF RESOURCE LIFETIME ESTIMATION

Equation 3.2 is integrated via analog programming as an example. As an exercise of this modeling, take a generic resource in its infancy, when the amount used in the first period is 10^{-3} of the T_o value, which we take to be 1. We take the growth constant to be 0.02, the total produced at any time is $U1$ (Equation 3.3), and the RHS can then be programmed as shown (Figure 3.4 (a)). Note that a simple VisSim© example is

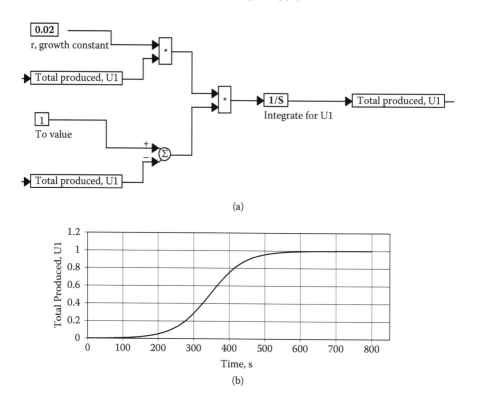

(a)

(b)

FIGURE 3.4 Analog solution and total production versus time for Equation 3.2, (a) program, (b) cumlative production.

given on the CD, CH3VIS, "Logistic model 1." The logistic growth of $U1$ toward a final value of T_o (i.e., 1) is plotted (Figure 3.4 (b)).

The value of $U1$ (amount produced) increases steeply at first, but then levels off, finally reaching the value T_o. At this point, and for whatever reason (exhaustion, change of tastes, a new more desirable resource), the value of $U1$ settles at T_o, the ultimate value. The production rate $dU1/dt$ is obtained before integration of Equation 3.3, and it peaks at the point in time corresponding to half of total production (Figure 3.5).

Of course, the ultimate T_o value of the resource may increase due to additional discoveries or to improved recovery technology. The multiplication of T_o is certain to be a matter of debate but removes a ceiling deemed questionable, given our state of knowledge of many resources. We assume, then, that T_o could increase in time as per the following equation:

$$T_o(t) = T_o \cdot MF \cdot (1 - \exp(-k1 \cdot t)) + T_o$$
(3.4)

In Equation 3.4, the original value of T_o grows toward $T_o \cdot (MF + 1)$ as time flows. MF stands for multiplication factor, and it represents one's estimate of how much the value of the resource might increase over the lifetime of the resource. The constant

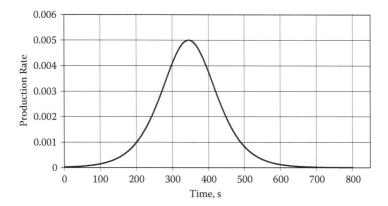

FIGURE 3.5 Production rates reach a maximum when half of the resource has been exhausted.

$k1$ simply reflects how fast the ultimate value is approached. A small $k1$ would mean slow and gradual discoveries, and a large $k1$ would reflect a few large discoveries early on. The inverse of $k1$ is a period of time in years, deemed to be the time interval for inventions to yield fruit. In analog programming, Equation 3.4 with an MF of 2 and a $k1$ value of 0.001 appears as in Figure 3.6, where "$T_o(t)$ value" stands for $T_o(t)$ in Equation 3.4. The resource varies over the simulation time as in Figure 3.7.

Hence, it is implicit that the resource doubles over the simulation time. The cumulative production $U2$ appears as in Figure 3.8.

Clearly, the slope (which is the extraction rate) decreases after 400 s even though the resource doubles. Such is the nature of exponential growth projections.

The growth rate can also experience random variations unrelated to the resource itself. Hence, a certain randomness can be reflected by adding a Gaussian random number of mean zero and standard deviation of 0.005 (Figure 3.9).

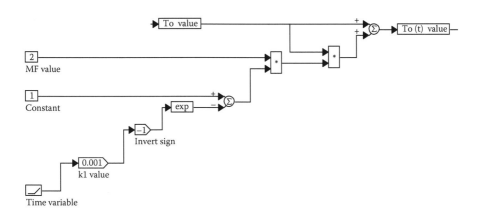

FIGURE 3.6 Programming of variable T_o.

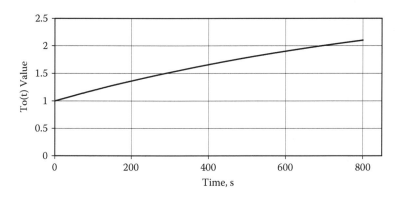

FIGURE 3.7 Temporal variation of the ultimate resource value.

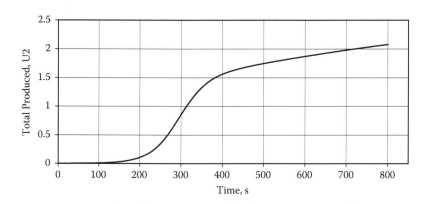

FIGURE 3.8 For resources under updating, the total produced is a function of time.

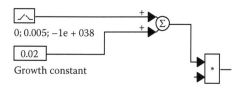

FIGURE 3.9 Addition of a random component to the growth rate.

Whereas the integration for $U2$ erases the randomness and results in the same trace as before, the production rate varies randomly. Production still peaks at a certain time, indicating that peaking production is built into Equation 3.3. It has not been proven that all natural resources follow Equation 3.3 in their growth and decline. Some resources, notably oil production within the United States, have reached a peak followed by a decline as of the present time.

PROJECTING INTO THE PAST

Application of Equation 3.3 calls for an estimate of the amount of the original resource that has been spent to date. This amount is the value of U at the present time (or at any time in the past). Whereas it is conceivable that such a database exists, it is not readily accessible for all fuels and all latitudes. In any case, an estimate, however coarse, is needed if one desires to explore the poignant peak question. If at the year that a resource is initially used its production is 1 (one), then

$$\frac{dU}{dt}\bigg]_o = 1$$

Under the right conditions its production will grow with time. At the end of the second year, its production will be

$$\frac{dU}{dt}\bigg]_1 = 1 \cdot (1 + gr) \tag{3.5}$$

where gr is the rate of growth. At the end of the third year, the production will be

$$\frac{dU}{dt}\bigg]_2 = \frac{dU}{dt}\bigg]_1 \cdot (1 + gr) = \frac{dU}{dt}\bigg] \cdot (1 + gr)^2$$

If a gr of 0.02 is assumed to hold constant over the time period under consideration, we then have the growth depicted in Figure 3.10, which shows that a consumption of 1 will grow sevenfold over 100 years given a gr of 2%.

Let us now invert the clock, making it run in reverse. Consider the production rate as of 2004 ($[dU/dt]_{2004}$). One year before (2003), the production can be estimated from Equation 3.5 as

$$\frac{dU}{dt}\bigg]_{2003} = \frac{\frac{dU}{dt}\bigg]_{2004}}{(1 + gr)}$$

and accordingly, over, say, n years, the production rate can be estimated as

$$\frac{dU}{dt}\bigg]_{2004-n} = \frac{\frac{dU}{dt}\bigg]_{2004}}{(1 + gr)^n} \tag{3.6}$$

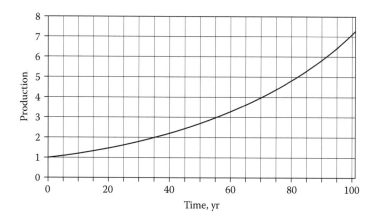

FIGURE 3.10 Compounded growth over 100 years.

We are interested in the sum of the yearly production since its initiation, namely the sum:

$$U]_{2004} = \frac{\dfrac{dU}{dt}\bigg]_{2004}}{(1+gr)^n} + \frac{\dfrac{dU}{dt}\bigg]_{2004}}{(1+gr)^{n-1}} + \cdots + \frac{\dfrac{dU}{dt}\bigg]_{2004}}{(1+gr)} + \frac{dU}{dt}\bigg]_{2004}$$

$$= \left[\frac{1}{(1+gr)^n} + \frac{1}{(1+gr)^{n-1}} + \cdots + \frac{1}{(1+gr)} + 1\right] \cdot \frac{dU}{dt}\bigg]_{2004} \tag{3.7}$$

The sum of the finite series of Equation 3.7 is easily computed by multiplying Equation 3.7 by $1/(1 + gr)$, which yields

$$\frac{U]_{2004}}{(1+gr)} = \left[\frac{1}{(1+gr)^{n+1}} + \frac{1}{(1+gr)^n} + \cdots + \frac{1}{(1+gr)^2} + \frac{1}{(1+gr)}\right] \cdot \frac{dU}{dt}\bigg]_{2004} \tag{3.8}$$

Subtracting Equation 3.8 from Equation 3.7, we obtain

$$\left(1 - \frac{1}{(1+gr)}\right) \cdot U]_{2004} = \left[1 - \frac{1}{(1+gr)^{n+1}}\right] \cdot \frac{dU}{dt}\bigg]_{2004} \tag{3.9}$$

Rearranging and simplifying, we obtain

$$U]_{2004} = \frac{\left[(1+gr) - \dfrac{1}{(1+gr)^n}\right] \cdot \dfrac{dU}{dt}\bigg]_{2004}}{gr} \tag{3.10}$$

So, given the production rate in any given year, the total consumed up to that year can be estimated if a suitable growth rate is known. Also, an average growth rate can be obtained if several production years are known. Of course, Equation 3.10 cannot replace actual data, but extrapolation into the past, when data are not available, can be attempted.

Example 3.2 studies Equation 3.10 and possible errors.

EXAMPLE 3.2 SUM OF TIME SERIES

Because consumption does follow the pattern leading to Equation 3.8, the sum of consumption is, in principle at least, easy to calculate. The difficulty lies in obtaining good values for gr, the growth rate (not to be confused with that used for the logistic equation). Many statistics point to gr for fuels of between 2 and 3% for the second half of the 20th century. However, variations for fuels and temporal periods exist, and careful consideration of average rates is a must. For our purposes we use estimated rates for a rough estimate of past consumption.

To gain an idea of the influence of different parameters on estimates, we compute the sum of the series using the closed form just derived:

$$U = \frac{dU}{dt} \cdot \left[\frac{1 + gr - \dfrac{1}{(1 + gr)^n}}{gr} \right]$$

for a given production rate today of $dU/dt = 10$ tonne/yr, and varying gr and n. The results are summarized in Table Example 3.2.1.

Clearly, the longer the past period under consideration, the greater the errors due to inaccuracies in the growth rate will become.

**TABLE EXAMPLE 3.2.1
Influence of Assumed Growth Rate
on Errors of U**

gr (%)	n (yr)	U (tonne)
0.02	10	99.8
0.03	10	95.3
0.02	20	173.0
0.03	20	159.0

UNITS: A PRACTICAL CHOICE

Whereas a preference for SI units exists in the field of engineering, the fact remains that most people can more easily visualize oil quantities. Hence, we adopt here the simple concept of toe, or tonne (2205 pounds [lb]), of oil equivalent (i.e., 7.33 barrels of oil) for energy. When the energy is electrical, we add an e at the end of the unit.

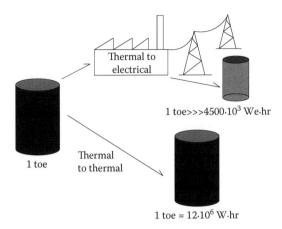

FIGURE 3.11 One toe in electrical and thermal terms.

For production rates, we use toe in one year, or toe/yr. For electric power production rates, we use We (Watts of electric power), but we convert the values to toee/yr. Electric energy is given in We·hr or toee. One unit of thermal energy cannot produce the same unit of electrical energy because conversions (cyclical or not) have efficiencies smaller than one, as explained in Chapter 1. Hence, we adopt an equivalence between thermal and electrical energy [2] that reflects the overall efficiency of this conversion:

$$10^6 \, \text{toe} \sim 4500 \cdot 10^9 \, \text{We} \cdot \text{hr}$$

which should not be confused with the straight conversion factor, namely

$$1 \, \text{toe} = 12 \cdot 10^6 \, \text{We} \cdot \text{hr}$$

The energy equivalence and the straight conversion are illustrated in Figure 3.11.

In order to gain a perspective on comparative resource magnitude and conversion efficiencies, in what remains of this chapter we adopt the units as tonnes of oil equivalent, toe, for energy, and for power, toe/yr. Table 3.1 summarizes unit values.

TABLE 3.1
Summary of Prefixes/Suffixes

Prefix/Suffix	Value
b	Billion, 10^9
M	Million, 10^6
e	Electrical

A RECAPITULATION

The simple approach to resource lifetime estimation is depicted with the aid of Figure 3.12. To project into the future, we use Equation 3.3 for logistic growth with calculated logistical growth rates r. To project into the past, we use Equation 3.10, with historical growth rates gr. The chosen units for production and cumulative production are toe and toe/yr (or its electrical equivalents, toee and toee/yr). The projection into the past is necessary to estimate U_o for use as the boundary condition in Equation 3.3.

THE FUELS

In what follows, we use the method outlined to speculate on the capability of fossil fuels to serve future needs. Included in the CD with this book, the reader may find VisSim examples for each fuel. To avoid redundancy, the examples are not described here, but they can be run with the viewer provided.

NATURAL GAS

Natural gas is a versatile, clean fuel that can be burned completely with little difficulty. More importantly, moving liquefied natural gas from source to consumption points is a rapidly emerging global trend. Natural gas is the fuel of choice for gas turbines, which, in combined cycle configuration, can reach efficiencies in the range of 55 to 60%. These are the highest efficiency ranges achievable with current commercial technology.

There is plenty of natural gas in the world, according to British Petroleum [2]. Let us consider proven reserves. As of the end of 2004, the proven reserves were

$$RW_{ng} = 163 \text{ b toe}$$

The estimated consumption as of 2004 (data are available from 1965 on [2]) was

$$UW_{ng} = 67.6 \text{ b toe}$$

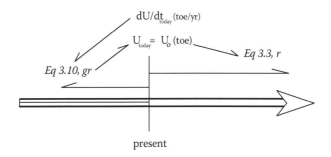

FIGURE 3.12 Projecting into the past and into the future.

We now need the value of r. We calculate it from Equation 3.3 so that, at the present time, yearly production equals 2.42 b toe/yr. Then,

$$rW_{ng} \sim 3.8 \cdot 10^{-4} \ 1/\text{btoe} \cdot \text{yr}$$

EXAMPLE 3.3 ESTIMATION OF r IN EQUATION 3.3

We start with Equation 3.3:

$$\text{Pr} = \frac{dU}{dt} = r \cdot U \cdot (T_o - U) \tag{3.3}$$

and the information given for natural gas in the world:

$$RW_{ng} = 163 \ \text{btoe}$$

The estimated consumption as of 2004 (data are available from 1965 on [2]) was

$$U = UW_{ng} = 67.6 \ \text{btoe}$$

$$\frac{dU}{dt} = 2.42 \ \frac{\text{btoe}}{\text{yr}}$$

$$T_o = RW_{ng} + UW_{ng} = 230.6 \ \text{btoe}$$

We now need the value of r. From Equation 3.3,

$$r = \frac{\dfrac{dU}{dt}}{U \cdot (T_o - U)} = 3.8 \cdot 10^{-4} \ \frac{1}{\text{btoe} \cdot \text{yr}}$$

We adopt this value for simulation.

The natural gas timeline (Figure 3.13) shows that the supply is ample. The curve levels off at 60 yr from now, and much can happen in that time span. This conclusion

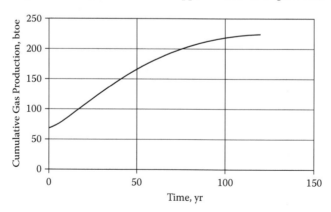

FIGURE 3.13 Cumulative production of natural gas versus time, logistic growth.

ignores the fact that gas is mostly available where it is not used, and that a substantial transportation system will be required to move the gas to points of use. See VisSim on the CD, CH3VIS, "Natural Gas Logistic model, World."

For the United States alone, matters are not that simple. We use EIA [5] data to obtain

$$RUSA_{ng} = 4.91 \text{ btoe}$$

We estimate the total consumed from Equation 3.10, starting with a production rate dU/dt for 2004 (0.47 btoe/yr), and decreasing each year by $(1 + gr)$ over 80 years. Then, for gr equal to 3% over 70 years, we obtain

$$UUSA_{ng} = 14.8 \text{ btoe}$$

We now need the value of r. This is difficult to estimate because, over the last 10 years the gas production has oscillated without a clearly defined trend. We again calculate r to match the yearly 2004 U.S. production from Equation 3.3, to obtain

$$rUSA_{ng} \sim 1.6 \cdot 10^{-4} \text{ } 1/(\text{btoe} \cdot \text{yr})$$

The cumulative production curve for the United States is shown in Figure 3.14. The natural gas outlook appears bleaker in the United States than in the world. The resource lifetime would appear to be 20 years. Our model is much too simple to encompass all the variables of importance, but it would appear that production of gas could decline in the future.

If the $RUSA_{ng}$ value were to change substantially, what would be the benefit of these additional discoveries? We can use the same model to gain an understanding of the impact of increased reserves. Let us postulate that, in a span of 50 years, we will double the value of $RUSA_{ng}$, according to Equation 3.4, with MF of 2. This MF

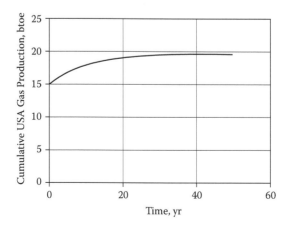

FIGURE 3.14 Cumulative production of natural gas in the United States.

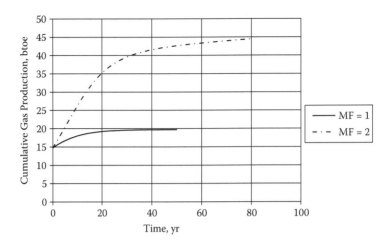

FIGURE 3.15 Cumulative production for variable $RUSA_{ng}$.

entails a favorable influence on the results, as shown in Figure 3.15. The curve does not level off for more than 50 years. See VisSim on the CD, CH3VIS, "Natural Gas Logistic model, USA."

So, the world natural gas supply is ample, but as for the United States, which, as of 2004 consumes 24% of the world's natural gas but produces only 20%, gas imports or fuel substitution will become more necessary. Something is driving the price of natural gas up, from $6 per 1000 cubic feet in 2002 to $15 as of this writing. It does seem that demand increases have outstripped new supplies, and that prices have consequently risen. Prices also tend to rise in the summer, reflecting increased air conditioning loads.

OIL

Oil, an elixir of radical importance for current transportation technology, is a mixture of molecules that are flammable liquids under ambient conditions. The high energy densities of liquid fuels from oil have enabled all types of liquid fuel-powered contrivances. The world situation regarding oil supply is a matter of much debate. The opinion of some [3] is that oil production peaked in 2005 and that we are now on a descending slope, the descent being irreversible. An equally convinced group [4] maintains that the best is yet to come, and that reserves are at least twice as large as estimated by those claiming a recent (or imminent) production peak. The ongoing discussion means that planning is necessary. As of 2004, we have for the world reserves [2]

$$RW_{oil} = 162 \text{ btoe}$$

How much oil has been extracted to this point in time we do not know, but doing the same extrapolation for the total production as we have done for natural gas, using Equation 3.10 for a consumption rate dU/dt (2004) of

$$dU/dt \text{ (2004)} = 4 \text{ btoe/yr}$$

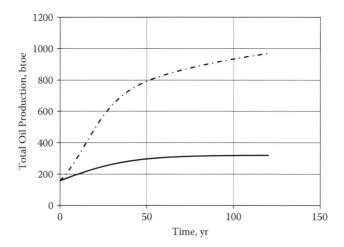

FIGURE 3.16 Cumulative oil production versus time for proven reserves (solid line), or proven reserves multiplied by 3 (dash- dot line).

and a 2.2% growth rate over 100 years, we obtain

$$UW_{oil} = 158 \text{ btoe}$$

We can adjust the growth constant to match yearly production to obtain

$$rW_{oil} \sim 1.6 \cdot 10^{-4} \text{ } 1/(\text{btoe} \cdot \text{yr})$$

which results in a cumulative production curve shown in Figure 3.16 (solid line). Of course, this logistic curve is not the end of the story because what if we could find three times the amount (possible as per [4]) over the next 100 years? Then the cumulative production curve would be as shown by the dash- dot line.

Clearly, the optimistic folk see no reason to worry. Production does not even level off for at least 100 years. The not-so-optimistic ones see the end clearly—50 years, at best, of declining production. Prudence would advise planning for the eventual demise of oil. See VisSim on the CD, CH3VIS, "Oil Logistic model, USA."

For the United States, if there is an optimistic chorus, it plays softly. If we take the reserves given by EIA [5] and extrapolate the 2004 production (0.248 btoe/yr) for 100 years at 2.5% *gr*, we get

$$RUSA_{oil} = 2.915 \text{ btoe}$$

$$UUSA_{oil} = 9.33 \text{ btoe}$$

$$rUSA_{oil} \sim 9.12 \cdot 10^{-3} \text{ } 1/(\text{btoe} \cdot \text{yr})$$

Using the logistical model shows that oil in the United States is precarious from a geological viewpoint [6]. Cumulative production levels off in approximately

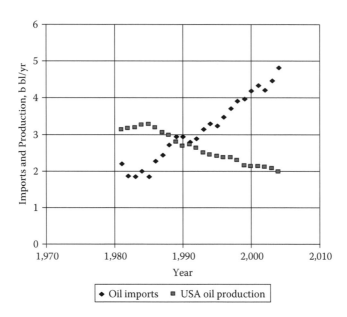

FIGURE 3.17 Oil imports into and production in the United States.

25 years, and doubling the reserves in 25 years (we repeat, there is no optimistic chorus) buys little additional time. See VisSim on the CD, CH3VIS, "Oil Logistic model, USA."

The United States imports well over 60% of the oil that it presently consumes. Data from EIA [7] show that imports have grown significantly, whereas production has declined steadily over the last 25 years. (Figure 3.17).

Prices reflect the volatility of international markets; in 2000 dollars, the price of oil, natural gas, and coal, according again to EIA [5], has varied sharply, and it continues to increase as of this writing. In ($/million BTU), the price has oscillated from $3 (1949) to $9 (1981) in constant 2000 dollars. In recent times, values of $20 have been recorded. Many factors influence the price of oil, especially when compared to that of coal, obtained almost completely domestically. As more oil is imported, it seems reasonable, if unpleasant, to expect further price fluctuations. Because gasoline demand has maintained an upward trend in the United States, whereas production has declined, it is fair to conclude that the availability of imported oil has closed the gap. By way of comparison, with the exception of solar energy, the world is currently (and for the foreseeable future) unable to import fossil energy from outside the planet, and it follows that a peak will occur, the only uncertainty being....when?

The lifetime of the oil resource from Equation 3.1 lacks the dynamic considerations of growth and depletion that we attempted to introduce with Equation 3.3. In terms of resource lifetime, certainly a less involved procedure, the size of the worldwide resource, is large:

World Li: ratio of 162/4, equal to 40 yr

and that of the United States is smaller, namely,

USA Li: ratio of 2.9/0.25, equal to 11 yr

The bleak oil situation in the United States could, under appropriate circumstances, elicit valid responses. Should we walk, ride the bus, buy a diesel or a hybrid, or an oil tanker (or shares in it)? All responses are possible, and some are much more effective than others in terms of reducing what we pay for energy. Yet, a large tanker may carry on the order of 2.5 million barrels of oil. Importing 5,005,541,000 barrels of oil per year to the United States [5] results in 2000 oil tankers per year docking at our shores, or about 6 tankers a day. Tankers are probably a good business for the future because, regardless of price fluctuations, oil is bound to supply a large percentage of the world energy needs as long as cars are available, and they constitute an important transportation mode. Only a concerted departure from this individual transportation mode would do away with the tanker business.

COAL

Whereas no fuel is really environmentally friendly with respect to production, conveyance, consumption, and by-products, coal can be a really difficult case. The average person's knowledge of underground mining is largely virtual, acquired through books, Hollywood, and the news. It has never been portrayed as a pleasant activity, and its danger to miners is constantly evident. Coal combustion can produce ashes and release minerals that we would do better to keep underground. That said, coal is rather abundant in the United States. The longevity of the resource could possibly be explained by the difficulty in handling it, but this speculation is far beyond our scope.

For coal, from the BP statistics for the world, we obtain

$$RW_{coal} = 462.6 \text{ btoe}$$

Determining the accumulated production to date is not simple. The 2005 EIA Energy Outlook [8] provides yearly production data from 1970 on. We calculate the average growth rate for that period, and with that rate, we extrapolate back to 1870 by using exponential growth. The cumulative consumption is determined by integration. The number thus obtained is bound to be less than accurate, but perhaps it is good enough for projections:

$$UW_{coal} = 102 \text{ btoe}$$

The growth constant $(5.7 \cdot 10^{-5} 1/(\text{btoe} \cdot \text{yr}))$ is chosen to match the 2004 production of 2.7 b toe/yr. With those data, we obtained the cumulative production curve in Figure 3.18. (See VisSim on the CD, CH3VIS, "Coal Logistic model, World".)

From Figure 3.18, the world coal supply seems ample for current production figures. This is the case for the United States as well. The total coal reserves (assuming 50% anthracite/bituminous and 50% subbituminous for conversion to toe) are [9]

$$RUSA_{coal} = 125 \text{ btoe}$$

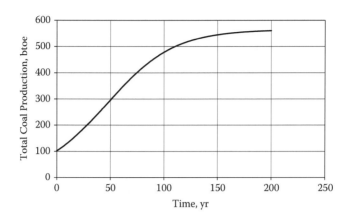

FIGURE 3.18 Cumulative projected production of coal over 200 years.

Using as total production to date the data in [10], we obtain

$$UUSA_{coal} = 39.2 \text{ btoe}$$

The growth rate to match the 2004 production figure (0.567 b toe) is $rUSA_{coal} \sim 0.00012$ 1/(btoe · yr). With these values, production levels off more than 200 years from now (see VisSim on the CD, CH3VIS, "Coal Logistic model, USA").

There is a small international market for coal, with stable prices. Hence, were it not for the release of CO_2 and harmful compounds (SO_2, Hg, NO_x, and other minerals), and the inherent dangers and difficulties of underground mining, coal would be an even more dominant source. It is clear that research on how to extract coal and how to use it more safely and cleanly could make a big difference in the endeavors of mankind. Such research is under way, but only in a half-hearted manner.

Uranium

Uranium serves as a fuel when its atoms are fissioned in nuclear power plants, generating various types of energy that are converted into heat. The heat is used to produce steam that runs turbines for power production. The sources of energy produced when an atom fissions can be either kinetic (i.e., that of the fission products and neutrons), or in the form of β particles and γ rays. Uranium is a most interesting energy source. Unlike fossil fuels, whose origin we can trace to solar energy via photosynthesis, uranium cannot even be produced by our Sun (it is not massive enough). Uranium is a sprinkle of the Universe that the Earth picked up in its forays through space. Yet, it harbors considerable energy, even if the by-products of its use are much more dangerous than carbon dioxide or NO_x at the same concentrations.

There seems to be some uncertainty as to how much uranium is available for use in current nuclear power plants. Estimation techniques, ranging from economic to net energy yield, produce estimates that vary widely. Because nuclear technology produces only electric power, we need equivalence between mined uranium and electrical energy produced. We will consider here that 190 tonne of mined uranium results, after processing and charging in a power plant, in 1 bWe · yr of electrical energy. Also, one tonne of mined uranium is equivalent to 10,252 toe devoted to produce electrical energy. (The rationale for these numbers is explained in Example 3.4.)

EXAMPLE 3.4 THE URANIUM RESOURCE IN TOE

We show here the conversion from tonne of mined uranium into toe. This conversion requires a number of assumptions about the way uranium is used, illustrated in Figure Example 3.4.1. This translation requires an efficiency in terms of mined tonne to energy produced. Such an aggregate number seems to range from 181 tonne/GW · yr to 200 tonne/GW · yr. Since the only large-scale use of uranium is to produce electricity, we then consider that 190 tonne of mined U can yield 1 GWe · yr.

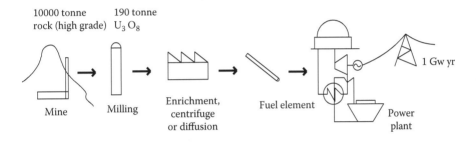

FIGURE EXAMPLE 3.4.1 From the ground to power.

It is necessary to estimate how many toe are required to produce the same amount of electrical energy. Because (Figure Example 3.4.2) we adopted the efficiency that 1 toe can produce 4,500,000 We · hr, then, we can produce a rough equivalence between 1 tonne of mined U and 1 toe as follows:

$$\frac{1}{\left(190 \cdot \dfrac{\text{tonne}}{\text{GWe} \cdot \text{yr}} \cdot 4500 \cdot 10^3 \cdot \dfrac{\text{We} \cdot \text{hr}}{\text{toe}}\right)} = 10252 \frac{\text{toe}}{\text{tonne}}$$

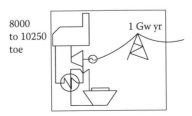

FIGURE EXAMPLE 3.4.2 From toe to Gw·yr via thermal conversion.

Thus, one needs 10,252 toe to produce the same energy in 1 tonne of U. This is quite remarkable. Factoring in other sources of data, it is found that the value ranges from 8,000 to 10,252 toe/tonne. Even when one considers the low concentrations of commercially extractable U oxide in the Earth, namely about 900 lb of rock per pound of U, one gets a ratio of 11 toe per ton of minable rock.

To estimate the current reserve for the world, the data of WNA [11] are used:

$$RW_{ur} = 36.2 \text{ btoe}$$

The total resource used to date cannot be ascertained accurately due to the dual use of the substance (i.e., power generation and weapons), but existing data [12] allow us to make an estimate of the total power produced by nuclear power plants until 2004, namely,

$$UW_{powerur} = 1.61 \cdot 10^4 \text{ GW} \cdot \text{yr}$$

To translate this number into tonnes of mined uranium, we convert the power to tonnes of mined uranium, and then to toe, to find the resource cumulative consumption of

$$UW_{ur} = 35.8 \text{ btoe}$$

We carry out our usual procedure (fit the growth constant to match yearly production of 0.697 btoe/yr) to obtain

$$rW_{ur} \sim 0.00062 \ 1/(\text{btoe} \cdot \text{yr})$$

which leads us to Figure 3.19.

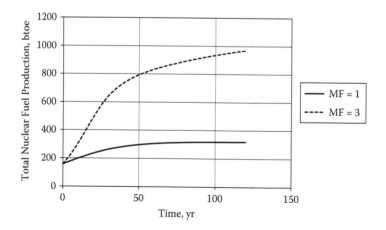

FIGURE 3.19 Cumulative production curve for mined uranium for the world.

The solid curve, if true, shows that, after 50 years, comparatively little additional uranium would be produced. To account for possible future developments, we assume that the resource will triple (MF = 3) over the next 100 years. Using this assumption, the dashed curve is obtained, which shows that cumulative production still grows after 120 years. Such resource multiplication is possible via advanced technology that still needs R&D.

The size of the reserves of uranium both in the world and in the United States depends on the price that utilities are willing to pay. According to EIA [13], less uranium is available at $66,000 per ton (2000 lb) than at $110,000 per ton.

$$RUSA_{url} = 1.20 \cdot 10^5 \text{ tonne at } \$66,000 \text{ ton.}$$

$$RUSA_{urh} = 4.04 \cdot 10^5 \text{ tonne at } \$110,000 \text{ ton.}$$

Or, in toe

$$RUSA_{url} = 1.2 \text{ btoe}$$

$$RUSA_{urh} = 4.2 \text{ btoe}$$

The higher number is adopted for calculations. It is even harder to determine how much of the domestic resource has been used as of today as once-through charge for the nuclear power plants. Summing up all the energy from nuclear sources produced from 1957 to 2004 gives

$$\text{Nuclear Energy} = 15.5 \cdot 10^{12} \text{ kW} \cdot \text{hr}$$

which, converted to mined uranium at 190 tonne/GW · yr and to toe, yields

$$UUSA_{url} = 3.38 \text{ btoe}$$

To check our approximations, we obtain another value of U by adding all the domestic production of uranium concentrate [14]. This second approach gives

$$UUSA_{ur2} = 4.32 \text{ btoe}$$

For our analysis, we adopt the average value of 3.38 and 4.32, namely,

$$UUSA_{ur} = 3.69 \text{ btoe}$$

A value of $rUSA_{ur}$ ~7.6 · 10^{-4} 1/(btoe · yr) is obtained for the high price, whereas the low price estimate yields $rUSA_{ur}$ ~2.6 · 10^{-4} 1/(btoe · yr). For each r value, the cumulative production curves are shown in Figure 3.20.

At lower prices, production levels off in 50 years, but more than 100 years' worth can be extracted at the higher price. Although these lifetimes are shorter than that of coal, they are certainly longer than that of oil or natural gas in the United States. Supply is not an immediate problem, then. (See VisSim on the CD, CH3VIS, "Nuclear Logistic model, World," and "Nuclear Logistic model, USA".) The difficulty with nuclear energy is the disposal of the wastes it generates, a topic that will be discussed in Chapter 8.

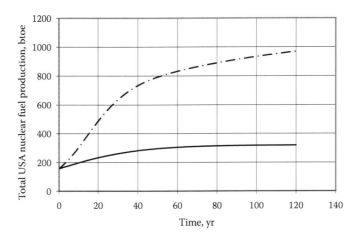

FIGURE 3.20 Projected cumulative values of mined uranium. Dash- dot trace corresponds to the high price, while the solid trace corresponds to the low price.

RENEWABLES

Renewables have been present throughout the history of mankind, particularly in the form of wind or biomass. Population grew at a steep rate as the Industrial Revolution unfolded (Figure 3.21) [15,16]. Experts debate whether energy availability and population are closely connected, but primary energy production and population show a close correlation.

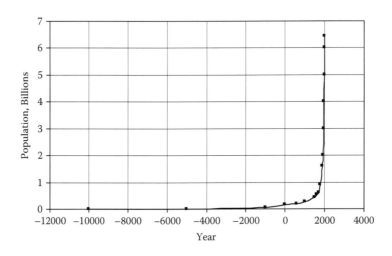

FIGURE 3.21 World population versus time.

Regardless of how evenly distributed energy use might be (and it is not, really), it is clear that, for renewables to fill a significant role, they will have to provide much more energy per capita, and much more total energy than in 1850. The question of our time is: can they? And if so, can they do this affordably? We address here only the first question. We focus on the gross potential of renewables in terms of resource magnitude. From a global viewpoint, renewables ceased to be the dominant energy source about 150 years ago. Many people in the world still rely on renewables for cooking or heating, so this statement should be interpreted in context. There are, for the industrialized world, energy sources that are more convenient, or cheaper than renewables.

We evaluate the resources of solar, wind, hydroelectric, and biomass. In addition, brief references to geothermal, tidal, and wave energy, are included. Rather than a logistical growth analysis, we focus on one of resource magnitude in toe/yr.

THE CENTER OF IT ALL

The Sun is the center of our solar system, even though it is nothing more than a relatively small star in one of 10^{11} (or more) galaxies in a Universe about whose boundaries we can only speculate. What matters to our endeavor is how much energy arrives at the Earth's surface, and what happens to this energy once it completes its long journey from the Sun in about 8 min. The fraction that is not reflected or absorbed into the atmosphere ends up striking the ground. Throughout the Midwest, the Sun bestows on the order of 5 kW · hr/(m² · day) as a year-round average. This means that about one barrel of oil (7.33 barrels equal 1 toe) lands on a 1m² surface (Figure 3.22) of Midwest ground over a year. Clearly, this energy can

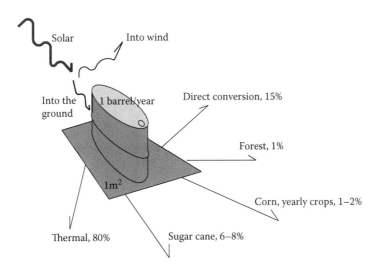

FIGURE 3.22 Some possible uses of solar energy, with approximate conversion efficiencies indicated.

be put to many uses, and some (such as the growing of sugarcane) belong in more tropical latitudes than the Midwest. However, these are the current possibilities around which solar can be evaluated: direct conversion via photovoltaic, thermal biomass, or wind, which arises when the incoming radiation heats the ground and thus the surrounding air. If the radiation falls on a body of water (ocean, river, or lake), then water vapor is generated. This vapor forms clouds and rain to sustain hydroelectric power generation.

In terms of efficiency, Figure 3.22 shows that direct solar conversion to electricity via photovoltaic is more efficient than conversion to biomass. The conversion to thermal may be attractive for both heating and power-producing installations. For the purposes of localized electrical power production, considering that solar cells can be mounted on roofs and that biomass cannot be easily subjected to the same procedure, direct conversion has clear yield advantages over biomass. This is even truer if the biomass conversion efficiencies are further reduced by processes required to convert biomass to usable forms such as electric power via combustion, or biofuels via fermentation and distillation.

SOLAR

The Sun, we state thankfully, shines on everyone everywhere, but only at *one time or another.* Clouds absorb and reflect solar energy, and are in constant motion. The lack of steadiness requires storage. Still, solar can be harvested to produce huge amounts of useful energy. How much? According to WEC [17], the Sun bestows $19 \cdot 10^{12}$ toe in one year on the land mass of the Earth, which covers about 29% of the world surface. Excluding the oceans, this would be a rate of about 25,200 TWe of high-grade energy. By high-grade we mean that this energy can be converted to electricity because it is electromagnetic radiation, not heat. The efficiency of direct conversion is not limited by Carnot, but assuming a relatively low overall conversion efficiency of only 5%, one could obtain electric power at a rate of 1200 TWe, or 900 times the rate presently generated by fossil and nuclear fuels.

Solar energy can be used thermally or for conversion to electricity. Conversion efficiencies to thermal can be high (from 60 to 90%), but the versatility of electricity makes the direct conversion especially useful. In the United States, the potential of solar is a function of the area considered and the technology implemented. Consider an average annual harvest of 5 kw · hr/m² · day, which seems reasonable, from the NREL map [18]. This is for a flat collector facing S and inclined at the latitude of the given location. Then, for an efficiency of 15% and using 10% of the area in the United States, we have

$$\frac{5 \cdot \text{kw} \cdot \text{hr}}{m^2 \cdot \text{day}} \cdot 9826630 \cdot \text{km}^2 \cdot 10^6 \cdot \left(\frac{m}{\text{km}} \right)^2 \cdot 0.1 \cdot 0.15 = 31 \text{ TWe}$$

This is indeed a large potential compared to the 0.41 TWe of fossil/nuclear plants. If one considers all the land area of the world, assumes a 5% efficiency for thermal collection and 2% for solar collection, and converts the resource to our adopted standard of toe, one obtains

World solar resource: thermal (5%) 950 btoe/yr, electrical (2%) 380 btoee/yr
USA solar resource: thermal (5%) 75 btoe/yr, electrical (2%) 30 btoee/yr

These values are coarse averages over wide areas, and certainly could bear refinement. Still, their order of magnitude is deemed correct, and it indicates a wealth of the resource.

WIND

The wind resource is, like solar, large. A recent study [19] evaluated the global wind resource by estimating the wind speeds at 80 m, a height suitable for 77-m turbines rated at 1.5 kW, for all continents. The technique and findings regarding wind speed were validated with data from a tower of suitable height. The wind resource was estimated at 72 TWe using 2000's technology. In 2003, conventional fossil fuel and nuclear plants generated power at a rate of 1.4 TWe globally. This does not include hydroelectric or renewable power. Hence, in terms of magnitude, the wind resource is sufficient to power today's world. However, winds are not always constant, and the populated areas are not always close to the generating areas.

In the United States, wind contributes power to the grid. Wind use is increasing rapidly. The maximum potential is land use and technology dependent, once sites with suitable wind speeds have been identified [20]. According to Ref. 20 for medium restrictions in land use and 50 m elevation (which is low for today's windmill technology), the potential would be 1.9 TWe. For severe land-use restrictions, the potential was estimated at 0.6 TWe. Because, as of 2004, fossil and nuclear power plants contribute 0.41 TWe to the national energy budget, there are good reasons to state that the contribution of wind could be significant.

As of 2005, installed wind capacity in the United States was 7 GWe. This is a small but fast-growing component of the total installed generating capacity of about 1000 GWe. Summarizing the wind resource in the chosen units, we have

World wind resource: electrical 140 btoee/yr
USA wind resource: electrical 3.7 btoee/yr

BIOMASS FOR FUEL

Biomass fuels have an element of outdoor roughness that many in developed areas strive to avoid except when camping. Yet, for a large number of people, these fuels are all there is. Biomass can be regarded as a way to store solar energy by chemical means. The problem is, the storage rate does not correspond to the rate of fossil fuel usage. In any case, estimates of the resource size differ widely. Reference 21 ventures a figure of 41 TW as an upper limit of the potential of this resource.

The U.S. potential for biomass is estimated to be somewhat low, on the order of 0.3 to 0.4 TW of thermal energy production rate, at $50/dry ton [22]. Energy plantations, dedicated to producing biomass for fuel, and using 10% of the available cropland, should result in an additional 0.1 TW production rate. Adopted biomass for fuel values are

World biomass resource: thermal 30 btoe/yr
USA biomass resource: thermal 0.4 btoe/yr

Biofuels

The topic of biofuels is a difficult one. The land for raising biofuels could be used for growing food instead. Also, as shown in Figure 3.22, the overall conversion efficiency is bound to be low, which leaves one wondering about the long-term prospects of this approach as population increases. Nevertheless, we present here some numbers which, if materialized, would affect the aforementioned estimates for solar, wind, and biomass. The rationale for biofuels is a strong one: liquid fuels pack a large energy density, and the whole transportation industry and its present infrastructure largely relies on engines that work well in conjunction with liquid fuels.

Biomass resources are difficult to estimate, but we have taken projections from the study in Reference 23. Many different scenarios were considered in this study, which makes interpretation of results difficult. Bounds (optimistic and pessimistic) in terms of technology and price can be found in the reference. We assume here a potential for a "middle of the road" future for biofuels in 2050, both for the World and the United States:

World: 4.2 btoe/yr
USA: 0.44 btoe/yr

Hydropower

As water evaporates from the oceans and lakes and then rains down on the mountains or high regions of the world, it gains elevation. As the water returns to the ocean, the elevation change can be used to run turbines. This is then a rather complex system, depending on fusion in the Sun for evaporation, on gravity (a fundamental force, the ultimate cause of which has not yet been established) for gaining elevation, on heat radiation to outer space for condensation, and on gravity again to power a turbine. For one of the oldest forms of energy production, this one depends on quite a bit that we do not fully understand. We rely again on the World Energy Council [24] for an ultimate limit for hydro of 1.6 TWe, comparable to what we generate today with fossil and nuclear.

Hydropower is the main contributor to power generation via renewables in the United States. Hydro energy produced in 2005 was 0.023 btoee [25], and the installed hydro capacity is similar to the nuclear capacity, or 100 GW. The consensus is that no significant additional resources are available for development in the United States, except in tribal lands. The use of those lands presents a number of conundrums that are beyond our scope. The resources are, then,

World hydro resource: 3.1 btoee/yr
USA hydro resource: 0.16 btoee/yr

OTHER SOURCES

Geothermal energy, tidal energy, and wave energy are other forms of renewable energy that never lose their appeal. In geothermal energy, heat from the Earth's core is used to generate power or for direct heat. Tidal energy uses the water level differences brought about by tides to run water turbines, and wave energy converts the motion of the waves to energy by a variety of proposed technologies. The World Energy Council site assessment of potential is sometimes vague, focusing more on what is (or could be) economical [26] than on what is possible. A potential of 0.85 TWe is established for geothermal power generation, whereas tidal and wave energy are assessed at 0.17 MWe each.

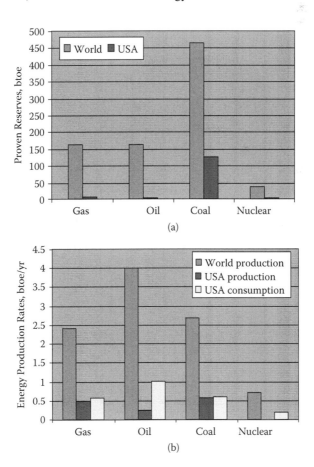

FIGURE 3.23 Selected reserves and production figures for 2004.

In the United States, geothermal energy generated 0.011 TWe in 2004 [27], with no significant tidal or wave energy reported. Projections of growth are uncommon, but EPRI [28] recently published a potential wave capacity of 20000 MWe. In the adopted units, this is

World geothermal resource: electrical 1.7 btoee/yr
USA geothermal resource: electrical 0.02 btoee/yr
World tidal and wave resource: electrical 0.6 btoee/yr
USA tidal and wave resource: electrical 0.04 btoee/yr

AN OVERVIEW: ALL TOGETHER NOW

Oil production has peaked in the United States in the early 1970s, but perhaps the reader will agree that timely peak prediction is a difficult, if not impossible, task. We present in this overview proven reserves and production data in comparative

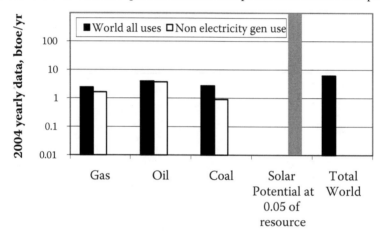

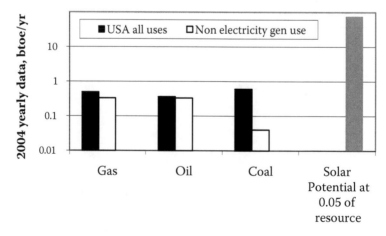

FIGURE 3.24 Comparison of fossil fuel production to 5% of the estimated solar resource (black series includes fuels for power production).

form, avoiding projections into the future. The intent is to foster meaningful comparisons among different energy sources to establish their potential as of the present time. Mankind consumes energy in basically two forms: thermal and electrical. Much of the thermal portion goes to produce electrical energy and mechanical work, and hence we will separate, to the extent possible, the thermal that goes to multiple uses and the thermal that is employed to generate electric/mechanical power. The data are those presented before for each resource. We converted all thermal units to toe to allow a certain perspective as to the potential magnitude of each resource.

Figure 3.23(a) shows the proven reserves both for the world and the United States. Clearly, the United States has major coal reserves. Regarding production and consumption (Figure 3.23(b)), as mentioned earlier, the United States is self sufficient only in coal. Much oil and some gas is imported to meet demand. Regarding the world, oil production dominates today as the major energy source.

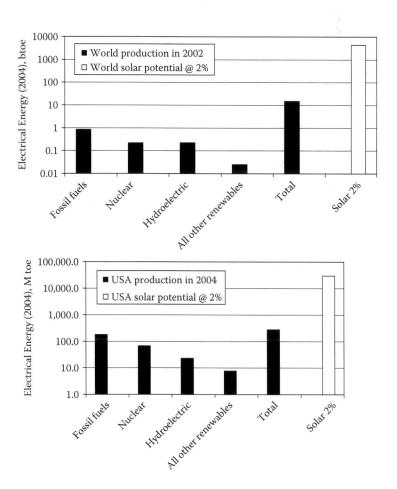

FIGURE 3.25 Recent electrical energy production and comparison to 2% of solar resource.

With respect to the solar resource for thermal purposes, one can say unequivocally that it is large (Figure 3.24). Even assuming that we only get to keep 0.05 of the incident energy every year, the resource is imposing, which does not necessarily mean that it is economical. Inevitably, developing it will require storage and backup systems of various kinds. The data for Figure 3.24 are the data presented in the previous section of the chapter. To compare only thermal data, the thermal energy devoted to power generation was deducted from the totals provided in this chapter. Hence, each graph has two series: in black, the total thermal energy; in white, the thermal energy not used for power production in utilities. Whereas coal contributes substantially to power production, gas does so to a limited extent (about 27% of the consumption generated power, even though the efficiencies achievable with gas are far superior to those achievable with coal), and oil contributes little. Power production uses are more noticeable in the U.S. scale than in the world scale, highlighting the high degree of electrification of the former. The reader must keep in mind the log scale of these graphs, which tend to visually enhance small numbers. Also, it must be borne in mind that oil is used primarily for transportation, for which solar cannot provide energy at the customary rates. The size of solar is nevertheless staggering, but so are the difficulties in harnessing it.

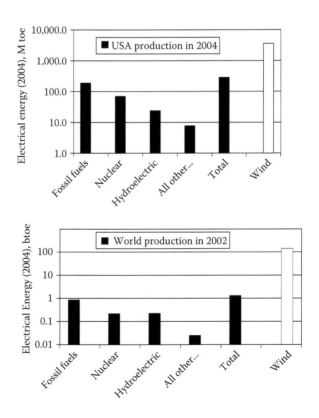

FIGURE 3.26 Recent electrical energy production compared to wind potential.

Electrical energy is most versatile, and it enables the best features of civilized life. Trite as this exhortation might be, imagine life without electric power. Hence, we make a direct comparison of the electrical energy produced in 2004 by different means to the potential of solar photovoltaic. Claims as to its affordability we make not; we are only after claims regarding its viability to provide for the human endeavor. In Figure 3.25, we see that, for the world, 2% of the solar resource for electrical purposes is ample. This presents the greatest engineering and financial challenge (other than wars) that mankind has probably witnessed: to harness all this energy for peaceful purposes. The challenge is no less significant for the United States. The current energy production is two orders of magnitude below that of the solar resource.

The wind potential compared to electric power production is shown in Figure 3.26.

REFERENCES

1. Smil, V., 2003, Against forecasting, *Energy at the Crossroads*, MIT Press, Cambridge, MA, chap II.
2. BP, British Petroleum, http://www.bp.com/multipleimagesection.do?categoryId=9010940&contentId=7021620
3. Deffeyes, K. S., 2005, *Beyond Oil: The View from Hubbert's Peak*. Hill and Wang, New York.
4. Wood, J., Long G., Morehouse F., April 2003. Long Term Oil Supply Scenarios, http://www.eia.doe.gov/pub/oil_gas/petroleum/feature_articles/2004/worldoilsupply/pdf/itwos04.pdf Also: April 2003 issue of *Offshore* under the title "World conventional oil supply expected to peak in 21st century."
5. EIA, Energy Information Agency, http://www.eia.doe.gov/oil_gas/petroleum/info_glance/petroleum.html
6. Hubbert, M. K., Nuclear Energy and Fossil Fuels, API Drilling and Production Practice, SP Meeting, San Antonio, TX, 1956, http://tonto.eia.doe.gov/dnav/ng/ng_enr_sum_dcu_NUS_a.htm
7. EIA, Energy Information Agency, http://tonto.eia.doe.gov/dnav/pet/hist/mttimus1m.htm
8. Energy Information Agency, http://www.eia.doe.gov/oiaf/aeo/pdf/trend_5.pdf
9. BP, British Petroleum, http://www.bp.com/liveassets/bp_internet/globalbp/globalbp_uk_english/reports_and_publications/statistical_energy_review_2008/STAGING/local_assets/downloads/pdf/coal_table_of_proved_coal_reserves_2008.pdf
10. Ristinen, R., Krasuhaar, J., 2006, *Energy and the Environment*, John Wiley & Sons, New York, Figure 2.7 data.
11. WNA,[a] World Nuclear Association,[a] http://www.world-nuclear.org/
12. BP, British Petroleum, http://www.bp.com/liveassets/bp_internet/globalbp/globalbp_uk_english/reports_and_publications/statistical_energy_review_2008/STAGING/local_assets/downloads/pdf/nuclear_table_of_nuclear_energy_consumption_2008.pdf
13. EIA, Energy Information Agency, http://www.eia.doe.gov/cneaf/nuclear/page/reserves/uresst.html
14. *Annual Energy Review 2004*, Report No. DOE/EIA-0384(2004) Posted: August 15, 2005 http://www.eia.doe.gov/emeu/aer/contents.html
15. CIA, Central Intelligence Agency, *The World Factbook 2005*, https://www.cia.gov/library/publications/the-world factbook/geos/xx.html

16. McEvedy, C., Jones, R., 1978, Atlas of World Population History, http://ehistory.osu.edu/world/

17. WEC[a], World Energy Council[a], http://www.worldenergy.org/wec-geis/publications/reports/ser/solar/solar.asp

18. NREL, National Renewable Energy Lab, http://www.nrel.gov/gis/images/us_pv_annual_may2004.jpg

19. Archer, C. L., Jacobson, M. Z., 2005, Evaluation of global wind power, *J. Geophys. Res.*, 110, D12110.

20. Elliott, D., Schwartz, M., 1993, Wind Energy Potential in the United States, September 1993. PNL-SA-23109. Richland, WA: Pacific Northwest Laboratory. NTIS no. DE94001667.

21. WEC, World Energy Council, http://www.worldenergy.org/wec-geis/publications/reports/ser/wood/wood.asp

22. ORNL[a], Oak Ridge National Lab http://bioenergy.ornl.gov/papers/misc/resource_estimates.html

23. De Vries, B., van Vuren, D., Hoogwijk, M., 2007, Renewable Energy Sources: Their Global Potential for the First Half of the 21st Century at a Global Level: An Integrated Approach, *Energy Policy*, Vol. 35, pp. 2590–2610.

24. WEC, World Energy Council, http://www.worldenergy.org/wec-geis/publications/reports/ser/hydro/hydro.asp

25. EIA, Energy Information Agency, http://www.eia.doe.gov/cneaf/electricity/epa/epat1p1.html

26. WEC, World Energy Council, http://www.worldenergy.org/wec-geis/publications/reports/ser/foreword.asp

27. EIA, Energy Information Agency, http://www.eia.doe.gov/cneaf/solar.renewables/page/geothermal/geothermal.html

28. EPRI, Electric Power Research Institute, http://www.epri.com/targetWhitePaperContent.asp?program=267825&value=04T084.0&objid=297213

4 Fossil Fuels and Their Technology

I thought of fire, but feared that the combustion of an infinite book would also be infinite, and would choke the planet with smoke.

J. L. Borges, *El Libro de Arena*

Engines that input thermal energy and output work are numerous. However, turbines (working with hot gas or steam) and internal combustion engines (IC, gas, and diesel) dominate the energy conversion scene of today. In this chapter, we describe these engines and associated cycles. Simple dynamic analyses of two fast-response engines (gas turbines and IC) are presented. Simulation programs are included on the CD.

NATURAL GAS

Natural gas is commonly a mixture of light hydrocarbons, but methane is the predominant component and the lightest. Being a combustible gas of small molecular size (Figure 4.1), methane diffuses fast, mixes readily with the oxygen in the air, and its combustion proceeds quite readily. Hence, high energy release rates are comparatively easy to obtain. In addition, methane is the feedstock for synthesis of a number of chemicals. The applications of those chemicals range from electronic equipment to fertilizers, including specialty gases and acids. Ideal combustion of methane, without excess air, produces water and carbon dioxide:

$$CH_4 + 2 \cdot (O_2 + 3.76\,N_2) \rightarrow 2\,H_2O + CO_2 + 7.52\,N_2$$

and seldom generates soot, resulting in a clear stream of combustion products.

Natural gas (other than as feedstock) can be used for direct heating or for power production. In direct heating, it can be used in furnaces to heat water or air as the heating medium. In power production, its combustion generates heat for conversion to electricity. Some natural gas is used for transportation in IC engines. Of all these applications, we select power production for further study. Of all possible conversions, power generation best uses the unique characteristics of natural gas.

GAS TURBINES AND COGENERATION

Gas turbines are a way of implementing the Brayton cycle. Following Figure 4.2, air is compressed (from 1 to 2), fuel is injected in a combustion chamber (from 2 to 3), and the hot gases are expanded in a turbine to release work (3 to 4). The turbine

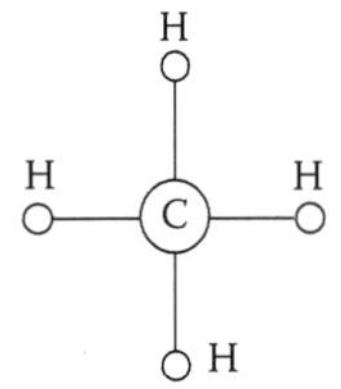

H
O

H   H
O——(C)——O

O H

FIGURE 4.1 Schematic
of methane molecule.

releases more work than required for compression, and
hence useful work is devoted to electricity generation.

The main idea behind the cycle is illustrated in the
schematic of Figure 4.3, where we simplify the cycle in
such a way that air is the working substance. Hence, the
combustion process is replaced with heat addition. The
chart shows the temperature versus the entropy of the air.
To any value of air temperature and entropy, there corresponds a point on the chart. When air is compressed
adiabatically (from 1 to 2), with entropy increasing due
to irreversibilities in the flow, point 2 lies to the right and above point 1 because the
compression operation increases the temperature, and the irreversibilities do so, too.
The temperature of constant pressure lines can be shown to increase (exponentially)
with entropy. At constant pressure, heat addition brings the temperature up to point
3, with T_3 being the highest cycle temperature, called, for short, the turbine inlet temperature (TIT). Adiabatic expansion in the turbine cools the fluid, still increasing its
entropy, to point 4. The reader may wonder why the compression and expansion are
adiabatic. The main reason is that these processes occur so fast that there is no time
for substantial heat transfer to occur.

The compression work (points 1–2) is smaller than the turbine work per unit air
mass, provided that the entropy increase in the compressor (or in the turbine) is not
overwhelming. As it was noted in Chapter 2, entropy increases in adiabatic systems destroy work, and hence they augment the work required by the compressor or
decrease the work available from the turbine.

EFFICIENCIES

The gas turbine efficiency is defined as follows

$$\eta = \frac{W_{net}}{\dot{m}_{fuel} \cdot LHV} \tag{4.1}$$

where *LHV* stands for the lower heating value of the fuel. Often, the inverse of
the efficiency is used, and this magnitude is the heat rate, typically in the units

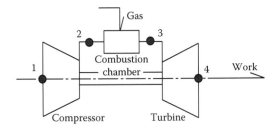

FIGURE 4.2 Schematic of a gas turbine.

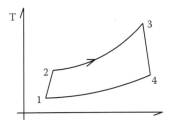

FIGURE 4.3 The Brayton cycle in the TS diagram for air.

BTU/kW·hr, which illustrate the ratio of heat input to work output. For small units, the percent efficiency ranges from the low to high 20s. For large gas turbines (100 MW or more), the percent efficiency reaches the high 30s. Fundamental thermodynamics show that the pressure ratio relates closely to the heat rate (i.e., to the inverse of the efficiency), a projection that is validated by practice (Figure 4.4). This figure correlates the pressure ratio to the heat rate of actual gas turbines.

The exhaust of a gas turbine is close to atmospheric pressure (point 4). Hence, for a given T_3, the temperature of point 4 is not completely independent, and typically falls between 600 and 850 K. Thus, heat can be recovered from the exhaust gases. These applications are called cogeneration or combined cycles. Typically, steam is generated from the exhaust stream and can be used for process, refrigeration, or power production.

When the steam is used to produce power, efficiencies in the 55 to 60% range are obtained. These are commercially available cycles, with the highest efficiency of all other commercial power-generating cycles. The basic principle of the cycle can be appraised with Figure 4.5.

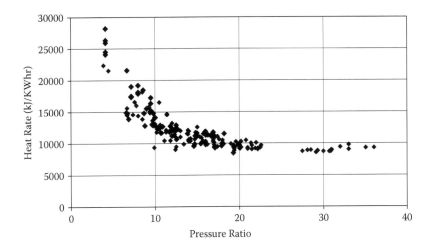

FIGURE 4.4 Heat rate versus pressure ratio for actual gas turbines.

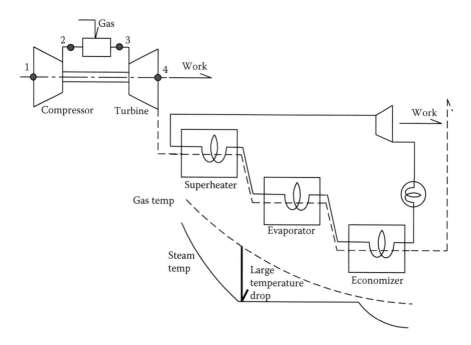

FIGURE 4.5 Cogeneration cycle with one steam pressure level.

To recover heat efficiently, counterflow heat transfer is thermodynamically the best option. As the gas temperature decreases after the turbine exit at point 4, it first superheats steam, then evaporates steam in the evaporator, and then preheats water in the economizer. The superheated steam flows through a turbine, producing more work. If an efficient gas turbine is used, then one unit of gas will produce about 0.38 units of work. The gas stream then has 0.62 units of thermal energy available, albeit not at the combustion temperature T_3 but at the much lower temperature T_4. If the stream is used to generate steam, as illustrated in Figure 4.5, there is a large temperature drop in the evaporator. This is a cause for irreversibility, and the combined cycle efficiency is on the order of 50%. To put some perspective to these numbers, consider that one pressure level enables recuperation of only 0.7 of the available heat. For a steam cycle efficiency of 0.33, the overall efficiency becomes

$$\eta_{1level} = 0.38 + (1 - 0.38) \cdot 0.7 \cdot 0.33 = 0.52$$

More pressure levels are used to approach reversibility, as shown in Figure 4.6. Three pressure levels reduce the temperature difference for heat transfer, thus decreasing the irreversibility. The gases can be cooled further, but three steam turbines are required. Because the gases can be cooled further, more heat can be recovered (we assume here that 92%, rather than 70% with one stage, can be recovered), and now the efficiency is higher:

$$\eta_{3level} = 0.38 + (1 - 0.38) \cdot 0.92 \cdot 0.33 = 0.57$$

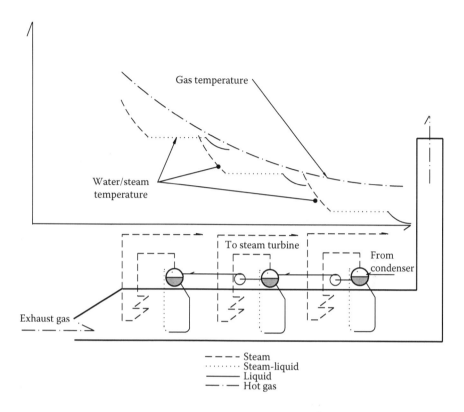

FIGURE 4.6 Heat recovery with three pressure levels of steam.

This is then the principle of cogeneration: to maintain the irreversibility at a minimum in the boiler to generate more steam at higher pressures.

BASIC OPERATIONAL PRINCIPLE

We explain here the principle of work extraction from the flow by a turbine of axial configuration. The turbine stage is composed of a row of nozzles and a row of blades, mounted on a disk. The nozzles accelerate a high-pressure flow, with the pressure decreasing as the velocity increases. The flow at high velocity impinges on the blades, imparting a torque and consequently a rotational motion to the disk. We show only one blade in Figure 4.7, mounted on a disk that is forced to rotate counterclockwise with velocity Ω as the flow impinges on the blades.

The blades have carefully developed profiles called airfoils. The flow approaches the moving blade with a velocity (relative to the blade) of W_1. In a properly working turbine, the flow goes around the blade and exits with a velocity W_2, also relative to the blade. If one could stand on top of the moving blade and visualize the flow velocities, one could record them as in Figure 4.8. When the two relative velocity vectors are brought together, one can see that the relative component in the

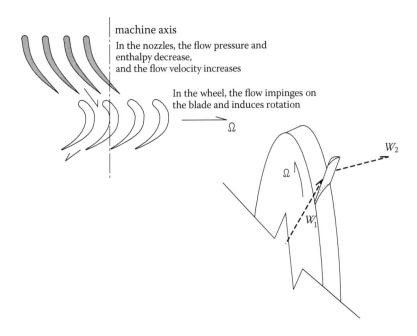

FIGURE 4.7 Schematic of turbine blade on disk.

circumferential direction $W_{1\theta}$ coincides with the direction of rotation. Coming out, the circumferential component $W_{2\theta}$ is opposite to the direction of rotation. The directional change is responsible for the change in the circumferential component. For the flow to change direction, some net force must act on it.

To investigate this action further, let us resort to our conservation laws. Consider a CV enclosing the blades. This would be a toroid of rectangular section, with inflow and outflow through the faces with normal parallel to the machine axis. Our CV is a moving one, namely, we still are moving at whizzing speed with the blades.

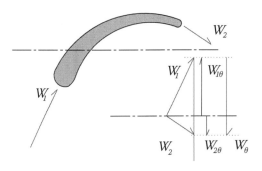

FIGURE 4.8 Directional changes of flow around a turbine blade.

According to Equation 2.9 (Chapter 2), we have

$$\sum \vec{F} = \frac{d\vec{p}_\sigma}{dt} + \dot{m}_o \cdot \vec{W}_o - \dot{m}_i \cdot \vec{W}_i \tag{2.9}$$

We consider forces only in the azimuthal (θ) direction. In the radial direction, we would have to contend with inertial forces necessary to explain our circular trajectory, but in the circumferential direction the noninertial nature of our coordinate system does not pose any additional forces, and hence we can write for steady state

$$\sum F_\theta = \dot{m} \cdot (W_{2\theta} - W_{1\theta}) = \dot{m} \cdot (\Delta W_\theta)$$

This equation shows that a net circumferential force is acting on the flow to deflect it. The direction of the force coincides with that of ΔW_θ (clockwise). The blades are exerting the force, and hence, by action and reaction, an equal force of opposite direction is acting on them. Thus, the force on the blades and the direction of rotation coincide, and the work is done on the turbine stage at a rate

$$\frac{dW}{dt}\bigg]_{stage} = Force \cdot Mean\ Radius \cdot \Omega = \sum F_\theta \cdot Rm \cdot \Omega = \dot{m} \cdot \Delta W_\theta \cdot Rm \cdot \Omega \tag{4.1a}$$

There are many turbine stages, and the work done on the shaft is the sum of the work done by each stage. If the distance between blades decreases, and if enough pressure is available, the flow will accelerate, and W_2 will increase. This effect is known as reaction. The turbine power, as per Equation 4.1a, will then increase, although stages with no reaction can be shown to provide higher power density under certain conditions.

The axial compressor also employs airfoil shapes to guide the flow into higher pressure states. However, the compressor airfoils are much different from those of a turbine. The flow is induced to turn by a rotating blade that receives power from the turbine. The force exerted by the flow on the blade is now opposite to the rotational direction, and work is done on the flow, as opposed to work being extracted from the flow as in the gas turbine.

TRANSIENT EQUATIONS FOR GAS TURBINE

The chief virtue of gas turbine engines for modern applications is their fast response to load variations. We formulate here some extremely simplified equations for studying the transient response of the turbine components. The equations are derived on the assumption that lumped parameters can be used to model the turbine. Since those lumped parameters are sometimes hard to determine, our modeling should be regarded as an approximation. Nevertheless, we will capture in our models the fast response times of the technology.

To retain the most realistic features of a gas turbine, we formulate a simple model based on the following premises:

1. As the engine speed increases, so does the flow rate. (Beyond a certain speed, the flow chokes and remains constant, but we will not endeavor to model this here.)
2. As the flow rate decreases, the compressor pressure ratio increases up to a limit.
3. Within the ranges under consideration, the equations relating the pressure ratio to the temperature ratio do not change.

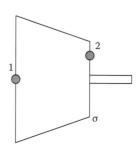

FIGURE 4.9 Compressor schematic.

Consider the compressor first (Figure 4.9). The First Law for the boundary σ around the compressor reads:

$$\frac{dQ}{dt} - \frac{dW_\sigma}{dt} = \frac{dE_\sigma}{dt} + \dot{m}_o \cdot \left(h + \frac{V^2}{2} + g \cdot z \right)_o - \dot{m}_i \cdot \left(h + \frac{V^2}{2} + g \cdot z \right)_i \qquad (2.14)$$

For an adiabatic compressor in the transient regime, Equation 2.14 (Chapter 2) becomes:

$$-\frac{dW_\sigma}{dt} = \frac{d\int_\sigma e \cdot \rho \cdot dV}{dt} + \dot{m} \cdot [h_o - h_i]$$

where it is assumed that no substantial accumulation of mass takes place. The integral signifies all the energy contained in σ, which we will take to be essentially thermal energy (although there are substantial amounts of kinetic energy in the flow). Then, for *CTM* equal to the compressor thermal inertia (the mass in σ times the average specific heat), we have, assuming for further simplification that the temperature is the average of the inlet and exit temperatures,

$$\frac{dE_\sigma}{dt} = \frac{d\int_\sigma e \cdot \rho \cdot dV}{dt} = CTM \cdot \frac{d(T_2 + T_1)}{2 \cdot dt}$$

A common simplification in compressor theory is to assume ideal gas behavior, namely,

$$(h_o - h_i) = c_p \cdot (T_2 - T_1)$$

With these simplifications, the energy balance becomes, considering that T_1 is constant,

$$-\frac{dW_{comp}}{dt} = CTM \cdot \frac{d(T_2)}{2 \cdot dt} + c_p \cdot \dot{m} \cdot (T_2 - T_1) \qquad (4.2)$$

The mass flow rate is taken (for the given inlet condition) as a function of the rotational speed as follows:

$$\dot{m} = -a1 \cdot \Omega^2 + b1 \cdot \Omega + c1 \qquad (4.3)$$

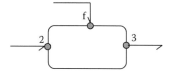

FIGURE 4.10 Schematic of combustion chamber.

The pressure ratio *pr* is also a function of the mass flow rate, as follows:

$$\frac{P_2}{P_1} = pr = -a2 \cdot \dot{m}^2 + \frac{b2}{\dot{m}} + c2 \tag{4.4}$$

The pressure ratio is related to the exit temperature via a polytropic relation

$$T_2 = pr^{pe} \cdot T_1 \tag{4.5}$$

where *pe* is assumed constant and a function of the isentropic efficiency of the compressor:

$$pe \sim 0.35 \tag{4.6}$$

A simplified version of the First Law for an adiabatic combustor (Figure 4.10) with no work done gives

$$0 = \frac{dE_\sigma}{dt} + \dot{m}_3 \cdot h_3 - \dot{m}_2 \cdot h_2 - \dot{m}_f \cdot h_f$$

Assuming ideal gas, a reference temperature T_o equal to 288 K, and a thermal inertia *BTM*, we obtain

$$0 = \frac{BTM \cdot d(T_3 + T_2)}{dt} + \dot{m}_3 \cdot cp_3 \cdot (T_3 - T_0) - \dot{m}_2 \cdot (T_2 - T_0) - \dot{m}_f \cdot LHV \tag{4.7}$$

The turbine responds to the same equations as the compressor, with a crucial difference: T_4 can vary as the TIT varies, or even as the pressure ratio varies. Accordingly, conservation of energy takes this form:

$$-\frac{dW_{turb}}{dt} = TTM \cdot \frac{d(T_3 + T_4)}{2 \cdot dt} + c_p \cdot \dot{m} \cdot (T_4 - T_3) \tag{4.8}$$

Assuming a negligible pressure drop in the combustion chamber, we can relate T_3 to T_4 as follows:

$$T_4 = \frac{T_3}{pr^{pe}} \tag{4.9}$$

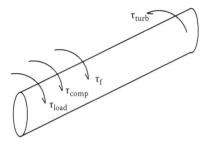

FIGURE 4.11 Actions on shaft.

where the exponent *pe* is assumed to be ~0.22. The key equation joining the turbine and compressor model is the dynamic equilibrium of the shaft. To present this equation, we note that, if the power *dW/dt* and rotational velocity Ω of a piece of equipment are known, then the applied (or output) torque τ is

$$\tau = \frac{\dfrac{dW}{dt}}{\Omega}$$

The torques acting on the shaft can be attributed to the turbine, to the compressor, to friction, and to the load. We call those actions τ_{turb}, τ_{comp}, τ_f, and τ_{load}, respectively. Hence, the shaft will have the following actions acting on it (Figure 4.11):

$$\tau_{turb} - \tau_{comp} - \tau_f - \tau_{load} = I_{turb} \cdot \frac{d\Omega}{dt} \qquad (4.10)$$

We now have enough equations to model the turbine if the fuel flow rate is specified. The reader can follow the following examples using the VisSim© viewer.

Example 4.1 (VisSim only on CD) develops a model using the foregoing equations to study gas turbine response to varying fuel inputs. When constant speed is necessary, as in power generation, the turbine model can be readily controlled as implemented in Example 4.2.

Example 4.2 (VisSim only on CD) shows how a proportional, integral (PI) control can be used to follow a varying load at constant rotational speed.

OIL

No other liquid in the history of mankind has endowed humans with so many fuels. Oil composition varies with location and even within a reservoir, and it must be processed to generate the various fuels and materials that make it currently of so much interest. When oil is subject to simple distillation, the lighter products come out at the top of the distillation tower, and heavier products further down. The number of carbons in the molecule determines the molecular weight of the compounds in question. The plethora of applications in Figure 4.12 is just the beginning. If one looks around the house and bus, train or car, one will find many objects, such as clothing,

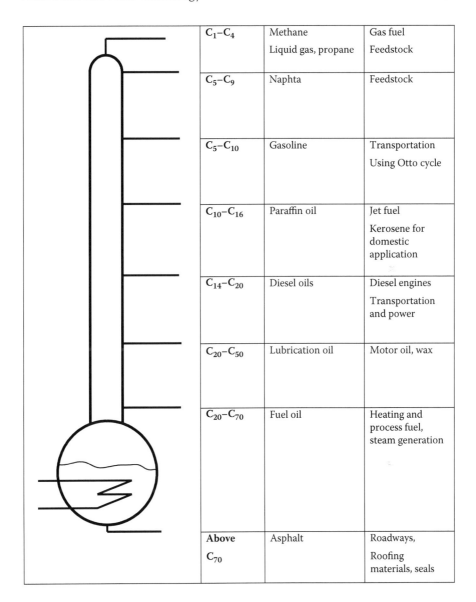

C_1–C_4	Methane	Gas fuel
	Liquid gas, propane	Feedstock
C_5–C_9	Naphta	Feedstock
C_5–C_{10}	Gasoline	Transportation
		Using Otto cycle
C_{10}–C_{16}	Paraffin oil	Jet fuel
		Kerosene for domestic application
C_{14}–C_{20}	Diesel oils	Diesel engines
		Transportation and power
C_{20}–C_{50}	Lubrication oil	Motor oil, wax
C_{20}–C_{70}	Fuel oil	Heating and process fuel, steam generation
Above C_{70}	Asphalt	Roadways, Roofing materials, seals

FIGURE 4.12 Oil and its many uses.

utensils, construction materials, and plastics, the manufacture of which require oil. Today, economies (of producer and consumer countries both) run on oil.

Because gasoline is clearly sought after more than other compounds, refiners use processes such as cracking to maximize the gasoline yield. Previous to cracking, vacuum distillation avoids the thermal breakdown of many heavy compounds, which may then be processed further to break their molecules down to make gasoline. In the United States, the product yield is tracked by the Energy Information

Agency (EIA). Whereas simple distillation will result in a 20 to 25% yield of molecules leading to gasoline, after downstream processing, the average gasoline yield is on the order of 45 to 50%.

The IC Engine

Whether for constructive or bellicose purposes, the idea of using the force arising from an explosion fascinated many for some 500 years before the 20th century. The past century saw the rise of this prime mover to a prominence never before achieved by any technology suited to the fulfillment of individual needs. It was the 19th century, however, that saw the first IC engines in modest numbers by Jean Lenoir in 1860, and particularly the four-stroke-cycle-based engine in 1862 by Nikolaus Otto. Siegfried Marcus (1870) put the first IC engine on a handcart, starting a saga that has not yet culminated—that of the automobile [1]. The manufacturing genius of the United States, coupled with its ample natural resources, allowed the creation of huge corporations devoted to individual transportation, with occasional, mostly short-lived, forays into buses or airplanes. Perhaps none of the IC engine inventors or entrepreneurs ever thought that their contribution would become as universal and controversial in terms of resource use and environmental stress as it has become.

What is then the appeal of the automobile engine? Other engines have been invented and built, but as of 2005, there were 500 million cars in the world, with slightly less than half in the United States, virtually all of them propelled by IC engines. Speed, or its illusion, is found exhilarating and not just by dogs sticking out their noses to the wind. A simple transient model might help explain the fast response to changes in load, a trademark of IC engines. To introduce the basic equations, we first focus on the principle whereby work is extracted from an expansion process.

Expansion Work and Temperature

The topic of conversion of heat into work via expansion is not hard to apprehend, but it has one subtlety first pointed out by Carnot: two sources at different temperatures are needed for cyclic operation (Chapter 2). Consider a primitive heat engine (Figure 4.13) devoted to lifting a weight F loaded on a frictionless piston of weight F_p. Most gases obey, to one degree or another, the following equation:

$$p \cdot V = Z \cdot Rg \cdot T \qquad (4.11)$$

Let us now start with the piston unloaded at p_1 and T_1. Then, if A is the piston top area, we have

$$p_1 = \frac{F_p}{A}$$

If we now move the weight F onto the piston, the pressure will increase to p_{1a}:

$$p_{1a} = \frac{F_p + F}{A}$$

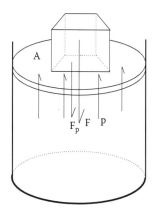

FIGURE 4.13 Simple piston engine.

Because the pressure has increased, the volume has decreased at constant temperature (Equation 4.11). Granted, the temperature could have increased momentarily, but we abandon here (for now) the intent to model transient phenomena and assume that eventually the temperature returns to T_1. We are now in a nice fix: we wanted to lift the weight, but it sank lower instead. We are not easily defeated, however, and, with an oil torch, we heat the cylinder and the gas inside it. As the temperature increases, so does the pressure (Equation 4.11). Lo and behold! with increasing gas pressure, the piston and the weight rise without us having to lift a finger, other than to light the torch, perhaps to get the fuel, or, certainly, to pay for it.

The pressure inside the cylinder is constant, but the volume increases with the temperature (Equation 4.11), and hence, the weight F rises. When we reach the desired height (say at a temperature T_2), we slide the weight off, and now, just like in some slapstick comedy, as we congratulate each other in triumph, the piston accelerates upwards, liberating the gas as it falls on the side. One can have too much of a good thing.

Indeed, if the external force does not correspond to that generated by the internal pressure, the piston will accelerate. We then refine the cycle: when the piston reaches the desired position, we lock it there, turn the torch off, move the weight out, and wait for the gas to cool down to T_1. We actually need to cool the gas down! As we unlock the piston, we now see it return to its original position, whereby we can load it with another weight and restart the cycle. Guess what? We need two heat reservoirs to pull this off: one a sink at T_1, the other a source at T_2. This conclusion is a general one: In the IC engine Otto cycle, net work is obtained only if two sources at different temperatures are available. Consider the four stroke cycle (Figure 4.14).

In the intake stroke, a piston moves, increasing the available volume while the inlet valve is open, and air–gasoline mixture enters the cylinder. When both valves are closed, the mixture is compressed and then ignited for combustion. Upon combustion, the mixture expands, doing useful work in the third stroke, turning the crank. In the last and fourth stroke, the mixture is swept out of the cylinder by the

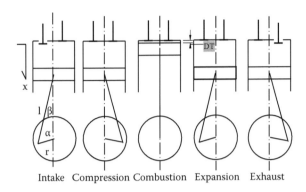

Intake Compression Combustion Expansion Exhaust

FIGURE 4.14 Otto cycle.

moving piston. In the PV diagram, the cycle appears as in Figure 4.15, drawn to show pressures and volumes inside the cylinder.

To attach a meaning to the indicator diagram, keep in mind Equation 1.8 (Chapter 1):

$$dW = p \cdot dV \tag{1.8}$$

This equation tells us that the area under a curve in the PV diagram represents work input ($dV < 0$) or output ($dV > 0$). Hence, if the compression (dash 2–3) curve coincided with the expansion (solid 4–5) curve, there would be no net work output because there

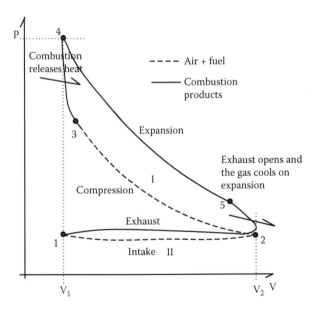

FIGURE 4.15 Otto cycle, schematic of indicator diagram.

would not be any difference in the areas under each curve. In addition, the intake and exhaust would require external work since the exhaust (5-1) is above the intake curve. Such a cycle would serve no purpose. Hence, expansion must proceed at higher pressure than compression. The only way this can happen is if the pressure increases from 3 to 4, and since a pressure increase at constant volume can only take place if the temperature increases (Equation 4.11), then combustion works as the "hot" heat source. The cold heat source is less evident; heat is removed by the engine coolant and carried away by the gases. When the expansion valve opens, say at 5, the sudden pressure decrease in the cylinder results in a gas temperature decrease, again, at constant volume. The thermal energy contained in the exhaust gases ends up in the atmosphere, as does the energy removed from the cylinder walls by the engine coolant.

IC engines have efficiencies that vary with speed. For well-tuned engines, efficiencies ranging from 0.2 to 0.3, are not uncommon, the former during transient-startup regime, the latter in steady state. Small engines with primitive valves and simple carburetors reach values on the order of 15%. Diesel engines, not covered here, tend to reach values in the 35 to 37% range.

SIMULATION OF ONE-CYLINDER ENGINE

The compression, combustion, and expansion strokes are unsteady processes that occur quite fast. Modeling of one cylinder retaining even the simplest features is quite an undertaking. We consider the geometry and thermal aspects of these transient processes. Figure 4.16 shows the nomenclature adopted to calculate the volume $Vol(t)$ versus time:

$$Vol(t) = (r + l + DT - r \cdot cos(\alpha) - l \cdot cos(\beta)) \cdot A_p \tag{4.12}$$

where

$$\alpha = \Omega \cdot t, \text{ and } \beta = asin\left(\frac{r}{l} \cdot sin(\alpha) \right)$$

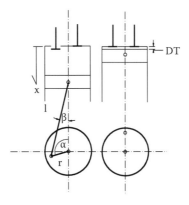

FIGURE 4.16 Geometrical parameters of the IC engine.

with Ω representing the angular velocity, and A_p the piston area. As the time variable t increases, the angles change periodically. At a given rotational velocity, the rate at which the volume in the cylinder is varying is given by taking the time derivative of Equation 4.12 to obtain

$$\frac{dVol(t)}{dt} = \left[-\Omega \cdot r \cdot sin(\alpha(t)) - \Omega \cdot \frac{r^2}{l} \cdot \frac{sin(\alpha(t)) \cdot cos(\alpha(t))}{\left[1 - \left(\frac{r}{l}\right)^2 \cdot [sin(\alpha(t))]^2\right]^{1/2}}\right] \cdot A_p \tag{4.13}$$

During the intake stroke, air and fuel enter the cylinder at a rate

$$\frac{dm(t)}{dt} = \rho \cdot \frac{dVol}{dt} \cdot \eta(\Omega) \tag{4.14}$$

where the volumetric efficiency $\eta_{vol}(\Omega)$ varies with speed, and we take, for example,

$$\eta_{vol}(\Omega) = 90 + 0.05 \cdot \Omega - 0.006 \cdot \Omega^2 \tag{4.15}$$

The compression process in an IC engine is complex because it happens very fast and gasoline evaporates during compression. We assume it here to respond to the following relationship between pressure p and volume V, with C2 a constant [2], the value of which is given in the VisSim example:

$$p = \frac{C2}{V^{1.33}} \tag{4.16}$$

Similarly for the temperature during compression,

$$T = \frac{C2T}{V^{0.33}} \tag{4.17}$$

The power for compression is given by

$$\frac{dW}{dt} = p \cdot \frac{dVol}{dt} \tag{4.18}$$

The energy balance in the combustion chamber is complex. Starting with Equation 2.14, we apply

$$\frac{dQ}{dt} - \frac{dW_\sigma}{dt} = \frac{dE_\sigma}{dt} + \dot{m}_o \cdot \left(h + \frac{V^2}{2} + g \cdot z\right)_o - \dot{m}_i \cdot \left(h + \frac{V^2}{2} + g \cdot z\right)_i \tag{2.14}$$

to the gas when the compression is nearly complete ($\dot{m}_o = \dot{m}_i = 0$) to obtain

$$\frac{dQ}{dt} - \frac{dW_\sigma}{dt} = \frac{dE_\sigma}{dt} \tag{a}$$

We consider the energy confined in the system to be the energy of the combustion products (gas, from now on), the energy embodied in the fuel, and the energy of the unreacted air:

$$E_\sigma = m_g \cdot c_{vg} \cdot (T_4 - T_o) + m_a \cdot c_{va} \cdot (T_3 - T_o) + m_f \cdot h_f$$

The temporal derivative of the foregoing equation requires care and interpretation:

$$\frac{dE_\sigma}{dt} = \frac{dm_g}{dt} \cdot c_{vg} \cdot (T_4 - T_o) + m_g \cdot c_{vg} \cdot \frac{dT_4}{dt} + \frac{dm_a}{dt} \cdot c_{va} \cdot (T_3 - T_o) + \frac{dm_f}{dt} \cdot h_f$$

Note, for instance, that whereas the total mass in the cylinder remains constant as combustion proceeds, the mass of gas, air, and fuel vary as the chemical reaction between air and fuel proceeds. From the immediately preceding equation and Equation (a), we have

$$\frac{dT_4}{dt} = \frac{\dfrac{dQ}{dt} - \dfrac{dW}{dt} - \dfrac{dm_g}{dt} \cdot c_{vg} \cdot (T_4 - T_o) - \dfrac{dm_a}{dt} \cdot c_{va} \cdot (T_3 - T_o) - \dfrac{dm_f}{dt} \cdot h_f}{m_g \cdot c_{vg}} \qquad (4.19)$$

where each derivative of mass must be used with its sign: dm_g/dt is positive because gas mass is increasing, whereas the fuel and air mass derivatives are negative because they disappear during the reaction. Equation 4.18 is modified to avoid large initial derivatives by increasing the heat capacity of the gas, which is initially zero:

$$\frac{dT_4}{dt} = \frac{\dfrac{dQ}{dt} - \dfrac{dW}{dt} - \dfrac{dm_g}{dt} \cdot c_{vg} \cdot (T_4 - T_o) - \dfrac{dm_a}{dt} \cdot c_{va} \cdot (T_3 - T_o) - \dfrac{dm_f}{dt} \cdot h_f}{m_g \cdot c_{vg} + m_a \cdot c_{va}} \qquad (4.20)$$

During expansion, the equations relating pressure, temperature, and volume are similar to those for the compression process (Equations 4.16 and 4.17):

$$p = \frac{C4}{V^{1.25}} \qquad (4.21)$$

The expansion power and work are also determined as per Equation 4.18 and its integral. The cycle efficiency is given by

$$\eta = \frac{Net\ work\ out}{Heat\ in} \qquad (4.22)$$

The heat input is calculated as

$$Q_j = LHV \cdot \frac{dm_f}{dt} \qquad (4.23)$$

Example 4.3 (VisSim on CD only) develops a model of one cylinder IC engine using the foregoing equations to study the influence of explosion timing on efficiency.

TABLE 4.1
Efficiency and Ignition Timing at Constant Engine Speed of 240 rad/s

Timing Advance (rad)	Efficiency
0.18	0.19
0.19	0.18
0.20	0.17
0.21	0.16

The transient nature of the cycle requires a timing selection mechanism. The TIMING BLOCK in Example 4.3 (on CD only) activates a unit signal for each stroke and allows the control of the combustion advance or the opening of the exhaust valve. The reader is encouraged to vary the ignition timing or the valve timing at constant speed to determine the efficiency response. Of note is the temporal variation of the fuel mass during combustion [2], which is assumed to respond, within the time allotted to combustion, as per Equation 4.24:

$$\frac{dm_f}{dt} = k \cdot m_f \cdot (m_{fo} - m_f) \tag{4.24}$$

where k is a suitable constant. Other representations of the fuel combustion rate have been advanced, notably that of Weibe [3]. The reader may recognize in Eq. 4.24 the logistical growth (in this case, decline) equation used in Chapter 3. In a way, it is remarkable that the same equation serves to describe such widely different phenomena. Other variables (inlet pressure, compression ratio, dimensions, and fuel mass) can also be changed. The influence of combustion timing on efficiency is shown, for this simple model, in Table 4.1.

COAL

Natural gas and hydrocarbons primarily have H-C and C-C bonds. The energy per C-H bond is 412 kJ/mol, and for C-C bonds 348 kJ/mol. Hence, an abundance of C-H bonds tends to result in higher energy release upon combustion. Octane, with 18 bonds, has a heat of combustion of 5471 kJ/kmol; propane has 2220 kJ/kmol; and carbon (graphite) has 394 kJ/kmol. Coal is composed of carbon and hydrocarbons, and many other inorganic compounds. Since it has comparatively few hydrogen bonds, its heating value per mole tends to be lower than that of hydrocarbons. Lower heating values and the solid (and soiled) state in which coal is found notwithstanding, it is a major player in power and steam generation. Coal serves as feedstock for a number of chemicals.

Coal is cheap and abundant, but it has three major drawbacks when compared to other fuels. First, it must be mined from surface or underground mines, which pose a variety of environmental problems as well as some dangers to miners. Second, it must be transported; and third, when burned, it releases not only the most carbon dioxide per unit of heat delivered but also sulfur oxides and other compounds that are not desirable. Nitrous oxides, common to all other fuels except nuclear, are present in coal. However, human ingenuity has contrived means of mining and burning coal that make it inexpensive enough to render it suitable for widespread use.

In increasing heating values per unit mass, coal can be found as lignite, subbituminous, bituminous, and anthracite. The heating values for anthracite range from 28,000 to 31,000 kJ/kg, compared to 43,000 kJ/kg for liquid fuels from oil, and 50,000 kJ/kg for natural gas. In addition, coal gasification and the viability of manufacturing liquid fuels from coal make it a very real possibility for continuing to be a strong contributor in the future. One example of coal gasification includes running the gases through a gas turbine. Coal gasification, however, requires water, and hence, an adequate source must always be available. Given that coal is largely used for power generation, we briefly describe here a conventional coal-fired power plant.

COAL-FIRED POWER GENERATION

Steam acquired relevance in the nineteenth century, when steam engines became the prime movers of the time, powered essentially by coal. Trains and steamboats were the buses, cars, and airplanes of the day, and factories were powered by steam. All those machines used expansion of steam to produce work (Figure 4.17). It was during this period that a distinguished, homeschooled engineer had the technical and business acumen to develop, introduce, and market steam turbines. Sir Charles Parsons indicates in a landmark lecture [4] how the idea of a turbine had been brewing for a number of years, and describes his contributions to the then-current state of the art.

The need to build ever-more powerful machines with increasing energy densities explains the success of steam turbines in replacing reciprocating steam engines. The early steam engines expanded steam in a piston (Figure 4.17). To do any work, the steam has to be at high pressure. In the return stroke, the steam is exhausted to the low-pressure condenser. The condenser pressure depends on the coolant temperature, and there is not much one can do about it. The maximum steam pressure depends on the flame temperature and on the boiler design, but again, metallurgical stability dictates maximum steam temperatures on the order of 820 K, and lower than 820 K in those early days. What is there to do if one wants more power?

Beyond increasing the steam pressure, the next option is to increase the area and rotational speed of the machine. This poses a problem: the piston must start, accelerate, decelerate, and change direction. Not much speed can be harnessed this way, as inertial forces (and associated stresses on connecting rods) grow with the mass or speed of the piston. Connecting several pistons together is a viable path (Figure 4.18), except that crankshaft design and durability has always been a challenge.

Clearly, as boilers could accommodate higher pressures, and condensers could operate at the subatmospheric pressures of saturated steam at room temperature, the stage was set for a more powerful machine, with a higher energy density. The steam turbine filled this role, and has been doing so for over a hundred years. The operational principle is similar to gas turbines (Figures 4.7 and 4.8). The steam accelerates

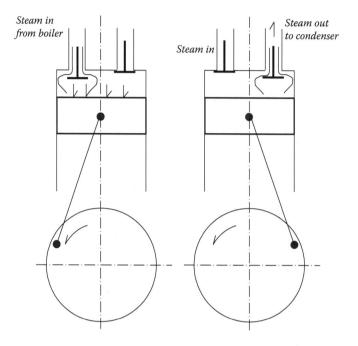

FIGURE 4.17 Schematic of one cylinder, single-acting steam engine.

before flowing between blades, which change the steam flow direction. When expansion occurs in the blades, the flow accelerates, and the turbine is called a reaction turbine. With no expansion, the concept is often referred to as impulse blading. Steam turbines (particularly the first stages) tend to be of the impulse type for a host of reasons beyond our scope, whereas gas turbines are of the reaction type.

Steam power plants are complex and evolving behemoths. In their fundamental unit, they comprise a boiler, a turbine, a condenser, pumps, and preheaters arranged so as to maximize efficiency and reliability. We follow the steam cycle with the aid

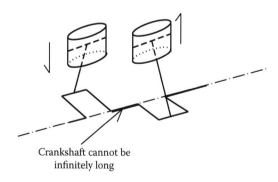

Crankshaft cannot be
infinitely long

FIGURE 4.18 Multicylinder concept.

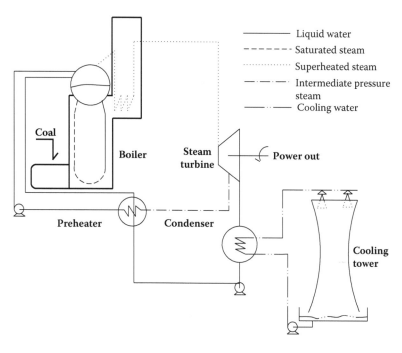

FIGURE 4.19 Schematic of a Rankine cycle.

of Figures 4.19 and 4.20. In the boiler, the top drum holds liquid water. The drum is connected to tubes and often pumps (not shown) to ensure circulation of the liquid inside the boiler tubes. The tubes are heated by the combustion of coal. The products of combustion flow in between the tubes and towards the stack via a superheater where the steam is superheated. The superheated steam flows through the turbine, producing power. Some steam is extracted at intermediate pressure to preheat the boiler feedwater. The remainder of the steam flows through the turbine to the condenser, cooled in this case with water from a cooling tower. The condensate returns to the boiler via the preheater (Figure 4.19).

The foregoing description is but a large simplification of reality, but (we hope) it retains its chief features. The intent of the component arrangement is to extract as much work as possible from the heat released by the coal before it goes up the stack. Envisioning this process is difficult, but we attempt to do so with a schematic Temperature–Entropy diagram for steam (Figure 4.20) where we superimpose the curves representing the gas temperatures and the cooling water temperature. This superimposition has only qualitative meaning in the horizontal direction since the entropies of steam and combustion products do not coincide. For the hot gas, in the vertical direction, it shows the gas cooling as the steam is generated in the boiler and then superheated. The steam then flows through the turbine and is condensed, transferring the heat to the cooling water. The cooling water is evaporatively cooled in a cooling tower. The condensed steam is preheated and returned to the boiler drum.

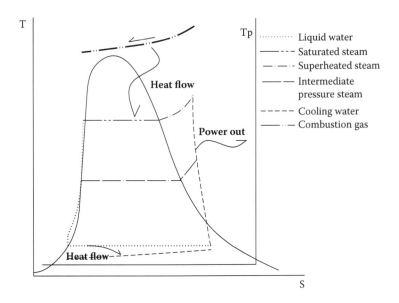

FIGURE 4.20 Schematic TS diagram for steam.

Typically, about 33% of the heat input is converted to work, substantially below a cogeneration system using a gas turbine (Figure 4.5).

The preheater raises the efficiency of the cycle by avoiding large temperature differences between the boiler feedwater and the hot gas. The cycle efficiency could be raised further by increasing the superheated steam temperature, but metallurgical limitations preclude this. A dynamic simulation of a coal-fired power plant is not offered here. In general, large coal-fired power plants do not respond as well to load variations as gas turbines do. The boiler and fuel thermal inertia are high, and hence, response times are long. Power plants of recent designs have Benson power generators, which by variations of the active area of the boiler can respond much faster than previous designs to load variations. Turbine capacities are large, up to or moderately exceeding 1 GW. Whereas a gas turbine compresses air before heat addition, steam cycles compress water of much smaller specific volume. Because the compression work is smaller for the water, many envision the Rankine cycle to have large efficiency potential.

NUCLEAR

In writing about energy technology, a decision must be made regarding whether to include nuclear energy in this chapter or in Chapter 5 for renewables. At least in theory, and explored to its ultimate possibilities, nuclear energy could be almost inexhaustible. This promising quality it shares with renewables. It is superior to renewables and comparable to fossil fuels in terms of the physical size of the power-generating plants, requiring only a fraction of the surface area called for by renewables. It is inferior to renewables in that the current fuel cycles generate long-lived waste that often has political ramifications, as addressed in Chapter 8. Nuclear energy cannot exist without governments providing a structure (physical and financial) to implement and safeguard both nuclear plants and their wastes. Nuclear plants are reasonably safe, nevertheless wastes are a symbol of the human trend to ignore the obvious fact that they are or could be dangerous. Curiously, this ignorance is also extended to the dangers of CO_2.

In any case, we describe the principles of nuclear technology and waste disposal from the perspective of an engineer who would like energy to be available to all of mankind so as to elevate the quality of life. Uranium is found in nature as isotopes $^{235}U_{92}$ and $^{238}U_{92}$, among others. It is a lucky find; our Sun (responsible for solar radiation, wind, and biomass) is not large enough to produce uranium. What there is of it seems to come from a supernova (or several) many eons ago. How the Earth picked it up is a mystery, but somehow the uranium is here, particularly in Australia, Canada, and Africa.

The mineral is present in the Earth's crust as oxides, of which the best known is U_3O_8. When purified from everything else, this oxide is called "yellow cake." It is sold to concentrating plants. In these plants, the oxide reacts with HF (hydrogen fluoride) to produce UF_4 in a kiln, and UF_6 after further reaction with gaseous fluorine in a fluidized bed [5]. Of the two isotopes, $^{235}U_{92}$ is prone to fission when bombarded with neutrons of the correct energy, and this isotope is the one desired for power plants. It is also the lighter isotope, and this difference in mass is used for separation.

There are two processes in use for separation. Gaseous diffusion is one of them. In this process, the UF_6 gas is compressed, and the lighter molecule (i.e., the one with $^{235}U_{92}$) diffuses faster than the heavier one through a membrane developed to this effect. Of course, many stages are necessary to reach the 3–5% proportion of $^{235}U_{92}$ used in some reactors. About 1400 stages are necessary, according to the World Nuclear Association [6]. The compression process consumes a lot of energy, which is rejected as waste heat. In the centrifuge process, the gas is fed to tubes that are rotating rapidly. The light hexafluoride migrates to the center, and the heavy one to the periphery of the tube. The rotational speeds are high, but the energy consumption, compared to the diffusion process, is small. Ten to twenty stages are necessary for proper enrichment.

The enriched uranium is reconverted to UO_2 and sintered into pellets that are fed into fuel rods. When the fuel rods are placed in assemblies and in a pressurized vessel filled with water, one has a pressurized water reactor. In this type of reactor, control rods absorb neutrons. The principle is as follows [7]: A few $^{235}U_{92}$ atoms will always split spontaneously. When they do so, a few neutrons are released. These neutrons do not have the necessary energy to collide and produce further fissions. A moderator (the metal of the rods, and the water) removes energy from the neutrons (i.e., the neutrons become "thermal neutrons") and renders them capable of initiating further fission reactions, if the control rods are withdrawn. A sketch of these possible sequences is shown in Figure 4.21.

Each fission releases energy in the form of radiation and kinetic energy of the fragments (the latter about 80% of the total). As a consequence of the absorbed radiation and the thermal energy arising from the conversion of kinetic energy into heat via friction, the fuel rod assembly heats up. When the reactor reaches steady

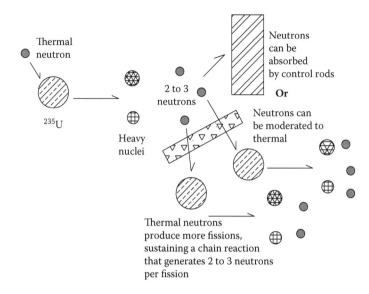

FIGURE 4.21 Fission and chain reaction.

state, the coolant is removing energy at the same rate as it is released by fission. The coolant flows to a boiler, where it generates steam, which runs through a turbine to produce power (Figure 4.22). The steam from the turbine is condensed and returned to the boiler. The pressurized coolant returns to the reactor via pumps.

A nuclear power plant rejects heat to the environment, just as the Second Law allows us to anticipate, and spent fuel must be removed every one or two years. The spent fuel is still radioactive, and from an energy accounting viewpoint is still rejecting heat via radiation and convection. Hence, the fuel utilization is not quite complete, and the efficiency could be higher. Several approaches have been proposed to deal with this "radwaste" [8]. More details on this issue are offered in Chapter 8.

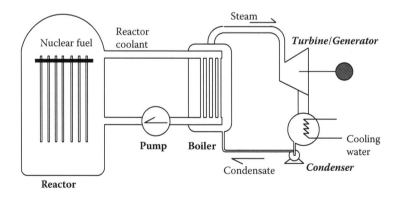

FIGURE 4.22 Pressurized water reactor schematic.

REFERENCES

1. http://en.wikipedia.org/wiki/Internal_combustion
2. Avallone, E.A., Baumeister, T., III, 1996, *Marks' Standard Handbook for Mechanical Engineers,* 10th ed., Chapter 9, McGraw-Hill, New York. Online version available at http://knovel.com/web/portal/browse/display?_EXT_KNOVEL_DISPLAY_ bookid=346&VerticalID=0
3. Ferguson, C., Kirkpatrick, A., 2001, *Internal Combustion Engines*, 2nd ed, p. 39, John Wiley & Sons, New York.
4. Parsons, C., 1911, The Steam Turbine, http://www.history.rochester.edu/steam/ parsons/index.html)
5. World Nuclear Association, 2008, The Nuclear Fuel Cycle, http://www.world-nuclear .org/info/inf03.html
6. World Nuclear Association, 2008, Uranium Enrichment, http://www.world-nuclear .org/info/inf28.html
7. World Nuclear Association, 2008, Nuclear Power Reactors, http://www.world- nuclear .org/info/inf32.html
8. Von Hippel, F., 2006, No hurry to recycle, *Mechanical Engineering*, Vol. 128, on the Web: http://www.memagazine.org/backissues/membersonly/may06/features/ nohurry/nohurry.html

5 Renewable Technology

It is worth exploring how solar energy actually works in all of its various manifestations.

H. Hayden, *The Solar Fraud*

For the energy scientist, renewables are a puzzle. Their possibilities are large, as seen in Chapter 3, but they have not attained widespread use in industrial societies. To unveil at least some aspects of this puzzle, we focus on proven technology: solar, wind, hydro, and biomass refuse combustion. This chapter does not exhaust the topic but focuses on what we deem of greatest potential for the world. Effective use of renewables probably constitutes the largest engineering challenge to be faced (thus far) by the species that placed humans on the moon, flew supersonically, built nuclear power plants, and crafted the amazing communications machine. Dynamic analysis of a solar collector and wind resource is illustrated.

SOLAR

Because the energy embodied in the Sun's electromagnetic waves is of high thermodynamic quality, enabling almost any transformation one could imagine, and because the energy itself is free, it is a most difficult topic to write about. It is difficult because, in spite of its availability and promise, solar energy has not "taken off," and many doubt that it ever will [1]. The problem, then, would be to explain why this energy source, obviously most desirable from all viewpoints, is not our dominant one. The explanation, if one exists, is so complex and involves so many interrelated disciplines and so many conflicting short-term interests that this writer cannot conceive how to begin, much less finish such explanation. Hence, this chapter focuses exclusively on an overview of available technology for solar, wind, and biomass without attempting to assess when or how renewables will acquire traction. That said, it is worth noting at this point that Sweden and Brazil run large segments of their economies on renewables, and that a handful of European countries have set their sights on renewables as a means to reduce CO_2 emissions.

Although free, solar energy is diffuse in two aspects: time and geography. When needed at night, it is not there. When needed during the day, cloud cover can decrease its availability. In winter, it is weaker than in summer in all locations except those along the Equator. Hence, solar implies storage or backup technology, which implies complexity and cost. Perhaps the only way to avoid independent storage (expensive, and with a negative impact on efficiency) is to derive baseline loads from conventional fuels.

Evaluating the solar resource in the United States is easy. The solar maps now available on the Web [2,3] lead to the conclusion that the following annual average irradiation is within reach of a considerable part of the United States (about 60% of the area):

$$PVW_{USA} = 5 \, \frac{kW \cdot hr}{m^2 \cdot day} \qquad (5.1)$$

About 40% of the area is above this value, and can reach up to 9 kW · hr/m² · day, and about 30% is much lower, as low as 2 kW · hr/m² · day. What this means is that 5 kW · hr/m² · day of energy (on the average) can be depended upon to impinge on one square meter in one day. The incoming energy can be converted into heat with maximum efficiencies in the 80% range, or directly into electricity, with laboratory efficiencies of about 40%. Consider now a few unit conversions and comparisons.

Example 5.1 shows how to estimate the energy arriving to the unit surface in barrels and toe, and also how much energy would be required to produce the arriving solar energy with known technology.

EXAMPLE 5.1 SOLAR ENERGY AVERAGE

Given the yearly average irradiation PVW_{USA}, determine its equivalence in barrels of oil per m², both in terms of thermal energy and power.

Given:

$$PVW_{USA} = 5 \cdot \frac{kW \cdot hr}{m^2 \cdot day}$$

Solution

a. Thermal equivalent
   On the average, the energy content of 1 toe is

$$1 \text{ toe} = 42 \text{ GJ}$$

About 7 barrels (br) make up one toe:

$$7.429 \text{ br} = 1 \text{ toe}$$

and with the energy equivalency

$$1 \text{ kW} \cdot \text{hr} = 3.6 \cdot 10^6 \text{ J}$$

we can conclude that, over one year, one square meter will receive

$$SE = 5 \cdot \frac{kW \cdot hr}{m^2 \cdot day} \cdot 365 \, \frac{day}{yr} \cdot \frac{3.6 \cdot 10^6 \cdot J/kW \cdot hr}{42 \cdot GJ/toe} \cdot 7.429 \cdot \frac{br}{toe} = 1.15 \, \frac{br}{yr \cdot m^2}$$

That is, one square meter (Figure Example 5.1.1) gets 1.15 br of oil equivalent per year.

b. Power

The power conversion factor reflects that oil needs to be burned in a power plant to produce power (Figure 3.11). Using the (Watt) We nomenclature for electric power and the equivalence explained in Chapter 3, then

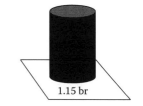

FIGURE EXAMPLE 5.1.1 1.15 br arrives over 1 m² in 1 year.

$$1 \text{ W} \cdot \text{hr} >>> 0.38 \text{ We} \cdot \text{hr}$$

and the solar input SEe in 1 year is

$$SEe = \frac{1.15}{0.38} \frac{\text{br}}{\text{yr} \cdot \text{m}^2} \sim 3 \frac{\text{br}}{\text{yr} \cdot \text{m}^2}$$

The interpretation of this result calls for utmost caution: this is an equivalence in that, to produce the 1.15 br of electrical energy would require, with today's technology, 3 br of thermal energy (Figure Example 5.1.2).

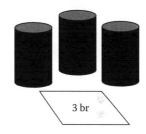

FIGURE EXAMPLE 5.1.2 To produce the solar electromagnetic energy arriving to 1 m² in one year would require, with today's technology, 3 br of oil.

How Big Is Small?

Small is beautiful, once the slogan for distributed energy, is seldom heard anymore. The average U.S. household consumes about 10,600 kW · hr in a year. The average roof size or distribution of a household is probably hard to give, and many households (apartments and condos) share one roof. Consider a hypothetical roof of a single household in the following example. Let us see how big and how efficient a solar PV system needs to be to meet average household demand.

Example 5.2 shows that household electric demand can be satisfied with relatively low efficiencies and collector size compatible with existing roofs.

EXAMPLE 5.2 AREA REQUIRED TO SERVE THE ELECTRIC LOAD OF ONE HOUSEHOLD

For roof dimensions L · W, calculate the solar energy striking it in 1 year. Compare this to the average electrical energy for household USA consumption.

Solution

For

$$PVW_{USA} = 5 \frac{\text{kW} \cdot \text{hr}}{\text{m}^2 \cdot \text{day}}$$

and taking as roof dimensions and area

$$Lrf = 25 \text{ ft} \qquad Wrf = 35 \text{ ft}$$

$$Arf = Lrf \cdot Wrf = 875 \text{ ft}^2 = 81.3 \text{ m}^2$$

we determine the energy received yearly by a collector facing south and with the inclination of the latitude corresponding to the location:

$$Erf = Arf \cdot PVW_{USA} = 81.3 \cdot \text{m}^2 \cdot \frac{5 \cdot \text{kW} \cdot \text{hr}}{\text{m}^2 \cdot \text{day}} \cdot 365 \cdot \frac{\text{day}}{\text{yr}} = 148354 \cdot \frac{\text{kW} \cdot \text{hr}}{\text{yr}}$$

Comparing the energy that can be obtained to the household average demand *HDe*,

$$HDe = 10600 \frac{\text{kW} \cdot \text{hr}}{\text{yr}}$$

it is possible to specify a required seasonal collection/storage efficiency:

$$\eta_{sol} = \frac{HDe}{Erf} = 7\%$$

That is, any efficiency exceeding 7% would satisfy the household electric power needs.

Since Example 5.2 shows that seasonal efficiencies of 7% are all that are required to serve a household, then, combined efficiencies of 15%, especially in areas of high irradiation, seem instrumental in making solar energy a reality.

YES, BUT WHAT ABOUT MONEY?

Many energy conversion technologies become viable because their development is funded by governments. The past century and our current time period overflow with examples of technology that became practical thanks to government leadership. Yet, no government is above physical law, which in practical terms means that it is impractical to subsidize energy sources when the resource itself is insufficient to sustain an economy. Solar is so abundant that it should not require subsidies, or in any event, it is abundant enough as to justify insightful subsidies to enable the ingenuity of engineers and entrepreneurs. The following example captures the savings that a homeowner could accrue with solar technology.

Example 5.3 discusses possible solar energy investments. Coarse numbers are shown to assess the possible investment for homeowners.

EXAMPLE 5.3 POSSIBLE SOLAR ENERGY INVESTMENTS

The average household consumption being 10600 kW · hr in a year, calculate the monthly average cost of electricity. Take $0.10/ kWe · hr as nominal price. Calculate the yearly cost for electric prices two and three times the given price and speculate as to how much capacity a homeowner could afford.

Solution

Household demand

$$HDe = 10600 \ \frac{kWe \cdot hr}{yr}$$

Cost of power

$$Cost = 0.1 \ \frac{\$}{kWe \cdot hr}$$

Yearly cost

$$Yearly \ Cost = Cost \cdot HDe = 0.1 \cdot \frac{\$}{kWe \cdot hr} \cdot 10600 \cdot \frac{kWe \cdot hr}{yr} = 1060 \ \frac{\$}{yr}$$

In 20 years, we save with a solar collector,

$$20 \ Year \ Cost = 1060 \cdot \frac{\$}{yr} \cdot 20 \ yr = \$21,200$$

In theory, one could spend $21,200 and break even if one had the funds available and wanted solar power. If the price of power doubled or tripled, then Table Example 5.3.1 shows the amounts per year or over 20 years that the owner could save with solar. For comparison, we include the cost of borrowing $25,000 over 20 years in the last column. Clearly, if the solar system cost were $25,000, one could borrow the funds and pay the bank or the utility about the same amount ($2,120) for power prices of 0.2 $/kW·hr. If the price of power exceeded the 0.2 $/kW·hr, or the price of the system were smaller than $25,000, then the owner has some serious paths to save money and enjoy an environmentally compliant system.

Owners about to experience increases of power cost could use Table Example 5.3.1 as a yardstick.

TABLE EXAMPLE 5.3.1
Higher Power Cost Makes Solar Attractive

Price of Power ($/kW·hr)	Yearly Cost of Power ($)	Power Cost Over 20 yr	Yearly Payment to Defray $25,000 Cost at 6% Over 20 yr
0.1	1,060	21,200	2,149
0.2	2,120	42,400	
0.3	3,180	63,600	

Regarding today's conditions, as shown by Example 5.3, it is fair to say that any technology aspiring to replace fossil electricity must reward the investment with savings amounting to a few years of energy cost. Escalating life-cycle fossil energy costs (including the cost of clean energy) may open a wider window for solar.

PHOTOVOLTAIC

We still have to address what one can do with the barrel of oil equivalent (or three barrels of oil for power production). Let us start with photovoltaic, for it entails the direct conversion from solar to electricity, and hence, each unit of electricity could displace up to three barrels of oil (or about a quarter ton of bituminous coal or 0.4 ton of lignite). Of course, an investment in photovoltaic would involve manufacturing energy and also some pollution that otherwise would not occur, but we must leave this for a more detailed analysis in Chapter 7, Table 7.10. A photovoltaic system configuration depends on whether the energy supplied can be DC or must be AC, and on whether storage is deemed necessary or not. In any case, we describe here a complete system, indicating, where appropriate, what components could be dispensed with.

The first word that comes to mind when looking at Figure 5.1 is "expensive," but let us continue. The PV cells do the rather amazing feat of transforming sunlight into DC power. The inverter transforms the DC power into AC power of the correct voltage and frequency, and may also route power to a storage system, typically batteries. The distribution panel routes the power to the load or to the grid, whereby the meter would run backwards, reducing the power bill.

PV Cells

Consider silicon (Si). On the Earth's crust, it is the second most abundant element (about 26% by mass), combined with oxygen in SiO_2, which we use abundantly to make glass. Quartzite, a form of SiO_2, is used to produce Si. After reduction by

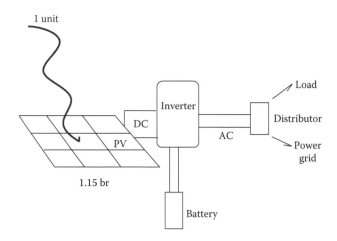

FIGURE 5.1 A PV system functional schematic.

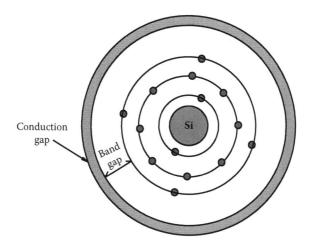

FIGURE 5.2 Schematic of Si atom.

carbon in an electric arc furnace, the Si is reacted with chlorine and reduced with hydrogen, resulting in pure Si. It can now be used for electronics. Recycled Si from electronics, or Si from silene (SiH_4) from the production process (decomposed by an arc), is used for solar cell production. This grade of Si is called amorphous Si, or Si with some hydrogen in it.

Si alone could be illuminated for eons without resulting in one microCoulomb going through any resistance. According to the atomic theory (simplified), Si has three main shells, with 2, 8, and 4 electrons (Figure 5.2). Each shell corresponds to a certain level of energy that decreases with distance from the nucleus. The four electrons in the outer shell are the valence ones, which form chemical covalent bonds with the other atoms. If light of sufficient frequency illuminates the atom, electrons can jump the band gap to access the conduction gap. Only if there is an electric field will the electrons move within the conduction gap. A semiconductor pair provides such a field in a rather surprising way.

Consider a Si crystal, where covalent bonds are formed with the four adjoining Si by sharing the four outer electrons. An n semiconductor (Figure 5.3) is obtained by replacing a Si atom with a P (phosphorus) atom. A p semiconductor is obtained when Si atoms are replaced with B (boron). The trick is that P, placed in the periodic table in the adjoining column to the right of Si, has an extra electron, and that B, from the left adjoining column, has one less electron. The role of Boron in modern technology is remarkable since it is also used in nuclear reactors to absorb neutrons.

The n crystal has extra electrons with relatively high energy, which makes it easy for them to reach the conduction band. The p crystal has holes, which can also change location, as the electrons shift around, filling them or leaving them empty. When an n crystal is placed adjacent to a p crystal (Figure 5.4), there is an interfacial region where the electrons and holes converge, with the electrons taking their place in the holes. The boundary is called a *junction*, and the region where the holes and electrons meet is the *depletion region*. This depletion region is neutral and acts as

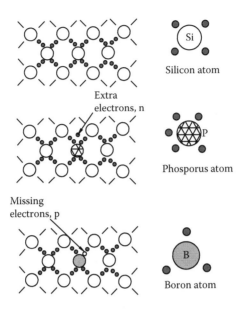

FIGURE 5.3 Si, n, and p crystals.

an insulator, preventing the migration of more electrons from n to p. However, an electric field builds up here since the n boundary adjacent to the depletion region has excess of holes, and the p boundary on the other side has an excess of electrons. The resultant electric field has the direction that a positive charge would follow; this is from n to p.

We have, then, a crystal with excess electrons that could fill holes if they could tunnel through the interface. This lack of equilibrium manifests itself as a voltage (on the order of ~1 V open circuit). When the extra electrons are collected with thin metal electrodes in n, they will flow to the p crystal, a process that unfortunately will soon cease as the electrons fill the available holes. Light can make this process last longer.

If the provision of electrons is steady, then they will flow from n to p continuously (Figure 5.4). Sunlight has photons of sufficient energy so as to raise electrons on both crystals to the conduction band. An excess of electrons is created in n (preferably in the depletion region). The electrons thus generated flow in the opposite direction to that of the electric field. The electric field is from n to p, and hence, the electrons flow to n. Those electrons then flow from n to p via an external circuit (with conventional current going from p to n). The power developed by the cell is the product of the voltage times the current available.

INVERTER

The inverter converts DC power into AC power. Hybrid automobiles depend on this technology by extracting DC power from batteries and converting it into AC power

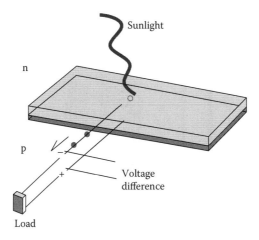

FIGURE 5.4 Sunlight generates a continuous voltage difference when striking a p–n junction.

of variable frequency. In a way, the solar energy inverters have an easier task, for the output is a fixed frequency. We explain here the basic aspects of inverter operation only (Figure 5.5). DC power enters a bridge that basically converts the DC into AC power, although the output wave may not be quite a sinusoidal one. The voltage of the AC wave is now adjusted via a transformer. Further filtering to remove the high harmonics and make the power compatible with the AC net power is done downstream of the transformer.

How to make AC from DC follows the basic principle illustrated in Figure 5.6. The load is connected to a DC source via four switches. The trick is now to open and close the switches in synch so as to subject the load to alternating polarity of AC power. To this end, imagine that S1 and S2 are closed, while S3 and S4 are open. The terminal A of the load will now be positive, whereas B will be negative. If S1 and S2 are now open, and S3 and S4 closed, as in the figure, then B will be positive, and A

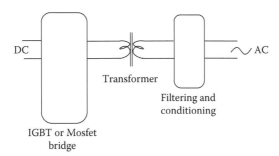

FIGURE 5.5 Inverter functional diagram.

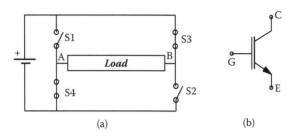

(a)   (b)

FIGURE 5.6 Inverter explanatory diagram, (a) functional schematic, (b) transistor switch.

negative. Converting the AC into DC calls, then, for four switches that can open and close at 60 Hz and do this continually with losses as small as possible.

Currently, the switching is done with a type of transistor recently made available, called the IGBT (insulated gate bipolar transistor). The construction of this power switch is complex and beyond our scope. The simple symbol is given in Figure 5.6(b). When the collector C is positive with respect to the emitter E, and the gate G is positive with respect to E, then the diode will conduct substantial amounts of power, up to levels on the order of 900 kW—quite a surprising power for electronics. Instead of the four switches of Figure 5.6(a), the transistors are installed in the bridge configuration of Figure 5.7. The gates must be activated and deactivated correctly at the appropriate time and with the correct frequency. Gate driver circuits exist for this purpose.

Inverter efficiencies, calculated as power output divided by power input, can be high. Reported values oscillate between 90 and 95%. For an installation connected to the grid, an inverter is necessary, but battery storage is not. For a completely isolated system, one could have batteries that store DC and dispense with the inverter. DC appliances are, however, hard to find.

BATTERIES

Batteries store energy in the form of chemical bonds. When a battery is charged, the energy furnished is stored in chemical compounds, the energy present in their bonds. Extracting the energy implies a chemical reaction at the expense of the compound that releases energy. As the energy in question is electrical, the means of input and

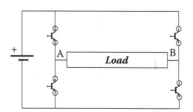

FIGURE 5.7 IGBT inverter bridge.

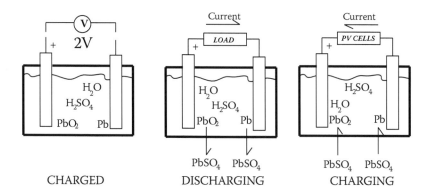

FIGURE 5.8 Lead–acid battery operations.

retrieval are currents flowing because of voltage differences. As the batteries have internal resistances, the Second Law indicates that energy will be converted from work into heat, and one never will retrieve as much energy as the input.

Lead–acid batteries, used for cars, are also recommended for PV systems due to cost and capacity reasons. We describe the way a lead–acid battery cell operates. In the charged state, the cell has aqueous sulfuric acid in a container with two electrodes—a lead dioxide electrode and a pure lead electrode. (A battery is composed of many cells such as the one described in order to provide a larger voltage than just one cell.) A voltmeter would measure 2 V for the fully charged cell (Figure 5.8). If a load is now connected to the wires, current flows from the lead oxide to the lead electrode. This means that electrons flow from the anode (lead) to the cathode (lead oxide). Then, the lead is oxidized, and reduction must take place in the cathode. In the anode, the lead is oxidized to give electrons:

$$Pb + SO_4^{-2} \rightarrow PbSO_4 + 2e^-$$

In the cathode, the electrons reduce hydrogen:

$$PbO_2 + SO_4^{-2} + 4H^+ + 4H_2O + 2e^- \rightarrow PbSO_4 + 6H_2O$$

As discharge proceeds, the electrodes become lead sulfate, and there is more water and less sulfuric acid in the solution. Eventually, the cell must be recharged by imposing a voltage greater than 2 V. The PV cells do this.

Battery efficiencies tend to range between 75 and 85%, as energy out divided by energy in. For our purposes, we select an efficiency of 82.5%. From Example 5.4, it does appear that the overall efficiency exceeds the minimum 0.07 required to satisfy the needs of the average household with a collector facing S. Nevertheless, reducing the average consumption via energy conservation measures should become a priority because the margins are not large.

Example 5.4 discusses overall photovoltaic efficiency. We project the efficiency of a PV system with and without storage.

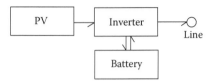

FIGURE EXAMPLE 5.4.1 Schematic of complete PV system.

EXAMPLE 5.4 OVERALL PHOTOVOLTAIC EFFICIENCY

Calculate the overall efficiency of a PV system (Figure Example 5.4.1). Compare to the minimum highlighted in Example 5.2.

For currently available equipment, one has

$$PV \text{ array } \eta pv = 0.15$$

$$\text{Inverter } \eta inv = 0.92$$

$$\text{Battery } \eta batt = 0.83$$

Solution

If the power is directly routed to the distribution network, the efficiency becomes

$$\eta pv\_line = \eta pv \cdot \eta inv = 0.14$$

If the energy is first stored, then released to the distribution network, one has

$$\eta pv\_batt\_line = \eta pv \cdot \eta inv \cdot \eta batt = 0.114$$

This value exceeds the efficiency calculated in Example 5.2.

WIND

Wind turbines are a way to harvest mechanical energy directly from the wind. Similarly to the Sun, the wind is seldom constant, and hence, storage may be required, unless the wind system is connected to a large grid served by fossil or nuclear. Some countries have staked their energy futures (or at least part of them) on the wind. Denmark has a prosperous wind industry that provides about 19% of the electricity demand. Wind resource evaluation is tricky because it depends on the assumed height and velocity profile for the wind. The wind velocity increases with H (Figure 5.9). The average velocity over the windmill area is V_w, and, clearly, it would depend on the assumed radius of the wind turbine.

Maps of wind resource abound, and for the purposes of our deliberations, we choose the following value, which represents well the Midwestern states:

$$WR_{USA} = 450 \ \frac{W}{m^2}$$

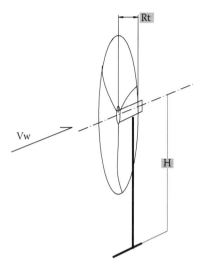

FIGURE 5.9 Windmill schematic.

One must be cautious with this number. It represents an average value of the energy in the wind. Design considerations for the wind turbine would call for the maximum energy value; storage considerations would call for the minimum energy value, as well as the duration of both maximum and minimum energy values. Also, there is a limit to how much of this energy can be harvested from the wind. The rest will be, well, gone with the wind. To understand this, we focus on the working principle of the wind turbine, which is essentially similar to the one presented for gas turbines in Chapter 4.

OPERATIONAL PRINCIPLE OF WIND TURBINES

The principle of energy extraction from a stream applies as in Chapter 4. Airfoils, if properly aligned with the stream, will extract power at a rate given by Equation 4.1a (Chapter 4). Figure 4.7 is reproduced as Figure 5.10.

As long as there is a change in relative velocity W_θ, the power extracted from the blade at the given radius Rm is given by (as shown in Chapter 4):

$$-\frac{dW_\sigma}{dt} = \dot{m} \cdot \Delta W_\theta \cdot Rm \cdot \Omega \qquad (4.1a)$$

for the blade rotating with velocity Ω in rad/s. So far, we have used only relative velocities, but to present the limitations of the wind, resort must be made to actual (absolute) velocities. The wind velocity in front of the windmill is $V1$ (Figure 5.11). Downstream, energy has been extracted from the wind, and the velocity is $V3$. The power extracted under steady-state conditions is, by virtue of Equation 2.14 (Chapter 2),

$$-\frac{dW_\sigma}{dt} = \dot{m} \cdot \left[\left(\frac{V3^2}{2} \right) - \left(\frac{V1^2}{2} \right) \right] \qquad (5.2)$$

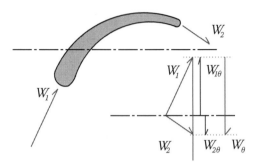

FIGURE 5.10 Directional changes of flow around a turbine blade.

The mass flow rate can be cast as a function of V2 as follows:

$$\dot{m} = \rho \cdot Aw \cdot V2$$

Taking

$$V2 \cong \frac{(V1 + V3)}{2}$$

we get that the power from the wind is

$$-\frac{dW_\sigma}{dt} = -P_w = +\rho \cdot A_w \cdot \frac{(V1 + V3)}{2} \cdot \left[\left(\frac{V3^2}{2} \right) - \left(\frac{V1^2}{2} \right) \right] \tag{5.3}$$

The maximum power from the stream would occur for V2 equal to V1 and V3 equal to zero, namely,

$$P_{w,\max} = +\rho \cdot A_w \cdot V1 \cdot \left[\left(\frac{V1^2}{2} \right) \right] \tag{5.4}$$

Dividing Equation 5.3 by Equation 5.4, we obtain the fraction of maximum power as cp, the power coefficient:

$$cp(V3, V1) = \frac{1}{2} \left(1 + \frac{V3}{V1} \right) \cdot \left(1 - \left(\frac{V3}{V1} \right)^2 \right) \tag{5.5}$$

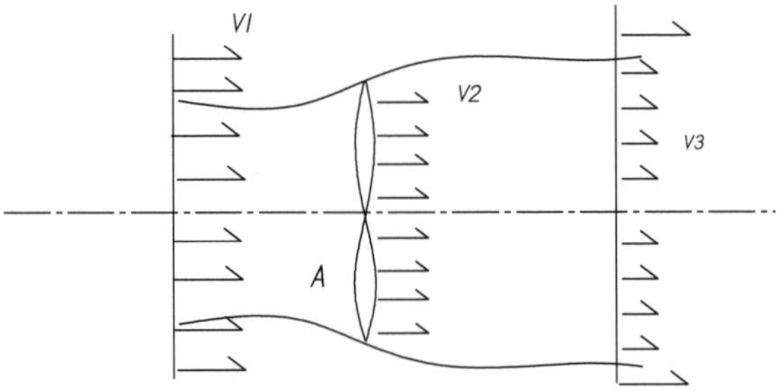

FIGURE 5.11 Wind velocities in a windmill.

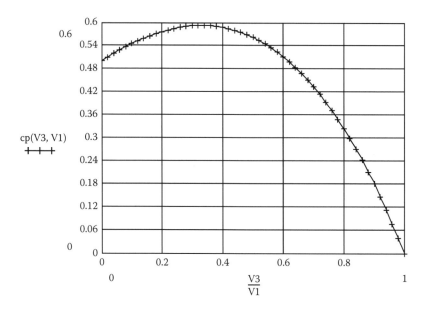

FIGURE 5.12 Velocity ratio to optimize power extraction from the wind.

This power coefficient is plotted in Figure 5.12. It is readily seen that the power extraction rate peaks at 0.59 for a velocity ratio of 0.35.

The actual efficiencies of wind turbines depend on the design, selected airfoil, and control system to keep the turbine and blades properly aligned with the relative wind direction for as long as possible. Also, turbines have different efficiencies (defined as energy output divided by energy extracted from the wind) at different speeds. For our purposes, we will assume an average efficiency of 25%. (This means a high efficiency on the order of 50% at the design wind speed.) We briefly describe the main components that make up a windmill—the rotor, the tower, and the generator. Controls and other details are left to specialized publications.

ROTOR

The rotor extracts kinetic energy from the wind using the principles of airfoil theory outlined in Chapter 4. The number of rotor blades depends on the design and intended applications of the windmill. Today, most windmills have three blades, although five and other odd numbers are possible in principle. According to Reference 4, three, as opposed to four, blades are preferred. This is because three-blade designs are more immune to vibrational excitation by the wind. In a three-blade configuration, as a blade experiences a solicitation when it is on top (Figure 5.13) (made up of a net force L in the direction of rotation, and a net force D in the wind direction), the other two blades, exposed to lower wind velocities, tend to compensate the solicitation in the D direction. With four blades, the blade opposite to the one on top is the low-velocity region adjacent to the tower. Hence, the D force is small, and the alternating solicitation is more severe for an even than for an odd number of blades.

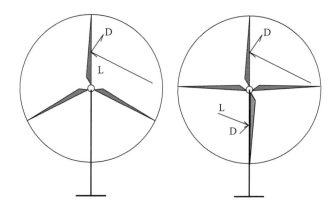

FIGURE 5.13 Three-blade versus four-blade designs.

The more blades a design has, the more costly it is, but the torque increases, and hence, at constant power, the rotational velocity can decrease. Diameters of wind turbines range today from 80 to 90 m, with blades up to 45 m made of composite materials. The rotor shaft is connected to a generator via a gearbox. The principles of generation via rotating machinery are common to all turbines, and are described in the following text.

Example 5.5 discusses power from the wind and dimensions, and we calculate the design and average power output of a windmill.

EXAMPLE 5.5 POWER FROM THE WIND AND DIMENSIONS

Calculate the rated power for a turbine of given diameter for the U.S. assumed resource for a design efficiency of 0.45 and an average yearly efficiency of 0.25.

Turbine diameter	$d_w = 85$ m	
Wind resource	$WR_{USA} = 450$ W/m^2	
Nacelle diameter	$d_w = 3$ m	
Efficiencies	$\eta ave = 0.25$	$\eta design = 0.45$

Solution

Power

At design $Pdesign = WR_{USA} \cdot \eta design \cdot \pi \cdot (d_w^2 - d_{na}^2)/4$ $Pdesign = 1.2$ MW

On the average $Pave = WR_{USA} \cdot \eta ave \cdot \pi \cdot (d_w^2 - d_{na}^2)/4$ $Pave = 0.6$ MW

Hence, whereas at design conditions the windmill would yield a very respectable 1.2 MW, as conditions change, the efficiency may not hold, and on the average, the power extracted could be lower than at design. However, the assumed case is deemed

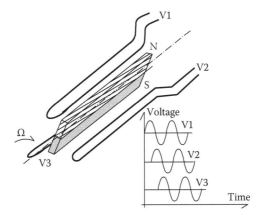

FIGURE 5.14 Generator schematic.

to be too extreme for a well-designed and controlled windmill, which can probably maintain an average efficiency much higher than the one quoted. Nevertheless, lower wind speeds reduce the energy harvested with the cube of the speed.

GENERATOR

The gearbox increases the rotational speed of the generator rotor to enable the line frequency. The basic principle of a three-phase generator is shown in Figure 5.14. The rotor is essentially a magnet with two poles. (Although the figure shows a permanent magnet, an electromagnet is commonly used instead.) Three distinct stationary coils surround the magnet. As the rotor turns, each coil faces a N and then a S. As the magnetic flux embraced by the coil varies, an electromotive force (emf) is induced, and an alternating voltage arises between the terminals of the coil. The next coil in the direction of rotation experiences the same flux in time but delayed by one-third of the rotational period. The next coil also produces AC but delayed two-thirds with respect to the first one. Each coil is then an AC phase, of which there are three in virtually every power-generating apparatus.

TOWER

The wind speed increases with distance from the ground. A tower capable of absorbing the solicitation of the rotor at all wind speeds must be designed. For large windmills, the tower height might be 60 to 100 m. Although a tower could be built as a lattice, the most common construction is a tubular steel column (Figure 5.15). The interactions between wind, turbine, and tower are complex. They could result in uncontrolled vibrations of rotor blades or towers. Consider for instance that wind gusts will increase the load on the tower because both the lift and drag will increase with wind speed (Figure 5.15(b), greatly exaggerated). If the wind gusts correspond to the eigenfrequency of the tower, a problem could result. Even if this is not the case (wind gusts are not regular), further excitation could result from the rotor blades, as one blade is partly unloaded as it moves in front of the tower.

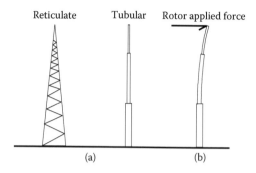

FIGURE 5.15 Wind towers.

Active damping of structures is a recent advance for vibration control that has been applied to some towers. In active damping, an actuator (or a series of them) applies a force to a mass connected to the structure via a spring and damper. As the mass accelerates, the structure is subject to the force transferred via the spring/damper. Vibration theory indicates that the transmitted force is not in phase with the applied one, but with a suitable control algorithm, the tower vibration can be reduced. Typically, the applied force is related to the velocity of the structure surface where a sensor is mounted.

AREA REQUIREMENTS

The area required by windmills is often the topic of debate. Hence, it is necessary to compute the area requirements. Large (MW) windmills have a small footprint, between ¼ and ½ acre. This includes auxiliary buildings, roads, and other requirements. The spacing between windmills is five to ten diameters, but this space is still open for other uses. A ratio sometimes ascribed to large wind farms is that they use about 2% of the total area. The rest can be used for crops or cattle. In Example 5.6, we use VisSim to establish how much energy can be obtained from a given wind speed distribution.

EXAMPLE 5.6 CHASING THE WIND WITH BATTERIES [VISSIM]

Although batteries are expensive and wind mills, for the most part, depend in the future on connections to existing grids, it is of interest to study what the required storage capacity to ensure a reliable service would be for a given wind distribution. The results are transferable to other renewables that share the random nature of the wind. We briefly describe the applicable equations, correlating them with the VisSim program.

For the period under study, we choose one week. The demand function for 1 week is assumed to adopt a simple periodic shape: 50% of the time the load follows a sinusoidal, as shown. The block that selects the sinusoid when positive is labeled "demand function." The details of this function can be best appraised by clicking on the block (Figure Example 5.6.1).

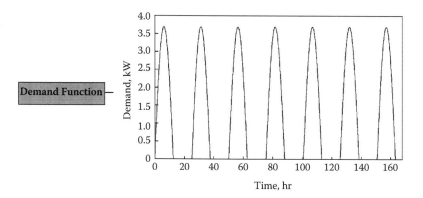

FIGURE EXAMPLE 5.6.1 Demand for the simulation period.

The wind follows a Weibull distribution with respect to velocity. This means that any given wind velocity at a certain location has a probability of taking place (over a long period of time) of

$$W2(vel) = \frac{\alpha}{\beta} \cdot \left(\frac{vel}{\beta}\right)^{\alpha-1} \cdot \exp\left(-\left[\frac{vel}{\beta}\right]^{\alpha}\right)$$

The Weibull distribution for

$$\alpha = 1.5 \qquad \beta = 5$$

has the following plot (Figure Example 5.6.2)

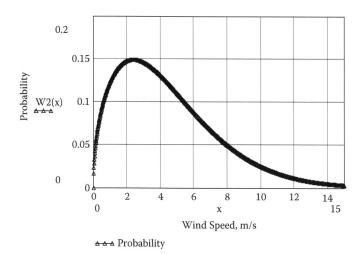

▲▲▲ Probability

FIGURE EXAMPLE 5.6.2 Wind velocity probability distribution.

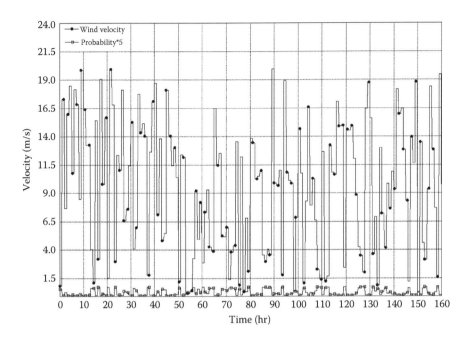

FIGURE EXAMPLE 5.6.3 Wind velocity (black dot) and its probability over the simulation period.

Although the mean velocity is 4.5 m/s, the mean power is higher than the one corresponding to this velocity because the power varies with the cube of the wind speed. Over the week of study, we assume that the wind changes every hour over a period of 168 hr or 7 days. The main difficulty lies in establishing how likely a certain wind speed will be in each of the hours of the week. In the block "Weibull Probability," the speeds are generated randomly, with an upper limit of 20 m/s. Each value, with a "Sample and Hold" block, is kept for one hour. A probability according to the Weibull function is assigned to each speed. The speed distribution versus time, with the corresponding probability is shown in Figure Example 5.6.3.

Clearly, high velocities have low probabilities. We assume a sophisticated wind mill design capable of extracting energy from all velocities, with the efficiency given in Figure Example 5.6.4. The block "Windmill Power" computes the power of the wind as per Equation 5.4:

$$P_{w,\max} = +\rho \cdot A_w \cdot V1 \cdot \left[\left(\frac{V1^2}{2} \right) \right]$$

The power is multiplied by the turbine efficiency and then by the probability of the velocity in that time interval (the Weibull probability times the total time interval, 168 hr in this case). The integrated power is compared to the total demand. If the amount of power harvested is insufficient to satisfy the kW · hr demand, the wind-mill

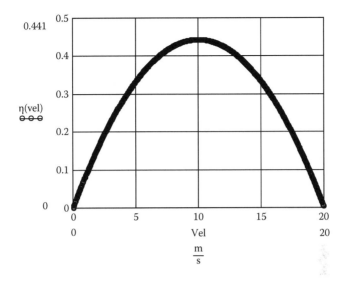

FIGURE EXAMPLE 5.6.4 Assumed efficiency of the wind turbine.

area is increased. Once the total power collected is the same or larger than the demand, we focus on the next block, namely, the battery storage.

The block "Battery Storage" compares the power production to demand. When production exceeds demand, the excess power is integrated over time to give energy, which is stored at an efficiency of 0.8. When demand exceeds production, the integrated difference is deducted from the battery charge. The battery charge in kW · hr is shown in Figure Example 5.6.5.

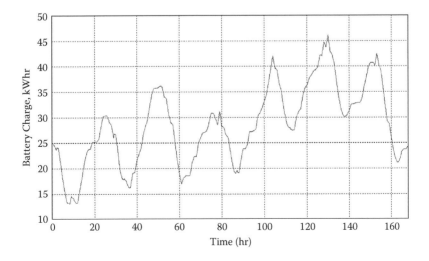

FIGURE EXAMPLE 5.6.5 State of charge of battery bank.

For this system, wind conditions, and windmill, it looks like a storage capacity of 46 kW · hr is necessary. A battery bank of 8 to 10 automotive type batteries is required. If the system starts the week at 25 kW · hr, it ends at 24 kW · hr, which is certainly adequate. Increasing the wind turbine diameter will harvest more energy. The extra energy could be routed to extra batteries, or, after conditioning, to the power grid. The latter situation, if generalized over the nation, will go a long way towards an environmentally friendly power system.

SOLAR THERMAL

As mentioned under the PV section in this chapter, solar electromagnetic radiation is a most versatile form of energy. It can be collected as electric energy, or it can be collected as thermal energy of virtually any temperature. We describe here the two technologies that allow the conversion from electromagnetic energy into thermal. The Second Law tells us that this is an "easy" conversion, for the conversion of work into heat is typically an effortless one. This is indeed the case here, albeit that once the heat is available, the challenge is to avoid its loss and to store it.

SOLAR COLLECTORS

There are essentially two types of solar collectors: flat plate and concentrating. The latter are capable of achieving higher temperatures than the former. We describe the flat collectors first. Solar radiation comes in several wavelengths. The energy per unit time, unit area, and unit wavelength interval is called the *irradiation*, and its dependence on wavelength is shown in Figure 5.16 [5]. The plot displays the energy

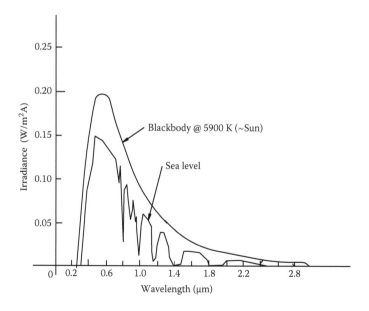

FIGURE 5.16 Spectral energy distribution for our Sun, approximated from data in Reference 5.

TABLE 5.1
Wavelength Ranges of Interest

Wavelength Range (µm)	Name
0.1 to 0.4	Ultraviolet (UV)
0.4 to 0.7	Visible (Vi)
> 0.7	Infrared (IR)

arriving to a surface normal to the radiation before and after it has gone through the atmosphere. After it has been through the atmosphere, energy has been extracted at some specific wavelengths. This is the energy absorbed by the gas molecules in the atmosphere, and, in a way, the absorption, as it warms up the gas, is partly responsible for the winds. At sea level or close to it, where most of us live, it is worthwhile to capture at least the energy present in wavelengths ranging from 0.3 to 1.4 µm. Of course, absorbing from 1.4 to 2.4 µm would be useful if one could keep that energy; however, it offers some difficulty in terms of radiation losses. In any case, the relevant wavelength ranges are listed in Table 5.1.

When the solar radiation impacts a surface (Figure 5.17), part of the energy is reflected, part is absorbed, and part can be transmitted. Each wavelength and direction may have a different behavior, but we assume surfaces gray and diffuse, for which we can dispense with directional and wavelength dependence.

Let us focus on an opaque surface, namely, with no transmission of energy. One can loosely correlate the temperature of a surface and the wavelength at which it will emit energy. From Figure 5.16, we can see that the Sun emits more or less like a black surface at 5900 K. When we approach a hot wall, such as that of a furnace or hot pipe, we can feel the radiation. The chair you sit on is cold, and there is no appreciable radiation coming from it. So, temperature and radiative power are loosely proportional to each other.

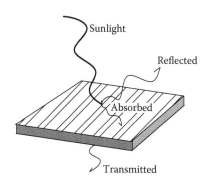

FIGURE 5.17 Solar energy can be absorbed, transmitted, or reflected.

For the purposes of a solar collector, the surface should absorb as much energy as possible in the shorter wavelengths that correspond to the amazing Sun temperature, and reject as little as possible at the larger wavelengths that correspond to its earthly temperature. We consider the Sun an amazing temperature because, as engineers, we find that no common materials can contain it: tungsten melts at 3680 K, whereas ceramics melt at about 3000 K. From a thermodynamic viewpoint, the radiation from the Sun is a bargain. In any case, a collector must then absorb the shorter wavelengths efficiently. The collector is at a much lower temperature than the Sun, typically 340 K.

At the lower collector temperatures, most emission of energy occurs in the wavelengths corresponding to 340 K, which are well in the IR range. Hence, the surface of collectors have selective coatings that absorb energy from short wavelengths but emit little in the IR range. Collector coatings typically absorb 0.9 to 0.95 of the incoming radiation in the visible wavelength range. In the IR range, they emit 0.1 of the energy radiated by a perfectly emitting surface.

CONFIGURATION OF A SOLAR COLLECTOR

A solar collector must absorb radiation, which is used to heat water. The design must be such as to route as much heat to the water as possible, avoiding losses to the atmosphere and substructure supporting the collector. In Figure 5.18 we show a schematic of a collector. The glass (low iron, to increase the transmission of short wavelength radiation) isolates the space around the tubes from the atmosphere. Hence, convection losses are minimized. The tubes are coated to maximize absorption. The plate supporting the tubes is also coated, and good thermal contact between tubes and plate must be ensured. The water circulating in the tubes takes up the thermal energy. Typically, the fluid is stored in an insulated tank. Hot water supply is the typical application of these collectors.

The efficiency of these collectors depends on their design, construction, and outdoor temperature. Unfortunately, the lower the outdoor temperature, the lower the

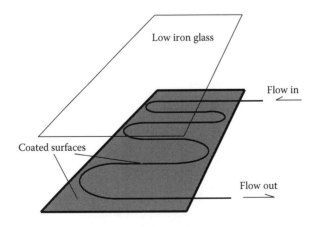

FIGURE 5.18 Flat plate solar collector.

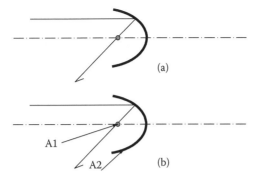

FIGURE 5.19 Parabolic concentrating mirrors.

efficiency will be. In summer, for domestic hot water, efficiencies of 60–70% are possible, which can decrease, for harsh winters, to 25 to 30%. Evacuating the space around the collecting tubes to minimize convection losses results in higher temperatures and in higher efficiencies in winter but lower efficiencies in summer: about 40% for the former, and 50% for the latter.

The nature of electromagnetic radiation is such that it can be concentrated to increase its temperature. Unlike a flat plate collector, the surface of a concentrating collector must be highly reflective and closely shaped to resemble a paraboloid. Those who paid attention to their high-school geometry will no doubt remember that every ray parallel to the axis will reflect in such a way as to go through the focus of the paraboloid (Figure 5.19). Hence, it is possible to concentrate virtually all the radiation incident upon the mirror in a relatively small area in the focus.

Consider the following equality that applies to the light flux for a small area A_2:

$$Ln_2 = \frac{A_1}{A_2} \cdot Ln_1$$

Hence, the energy flux (Ln, W/m²) of the incident light is multiplied by the area ratio to obtain the flux incident on the focal area. Concentrating collectors can achieve, by means of concentration, high temperatures. Those who did not pay any attention to paraboloids in high school can no doubt remember igniting papers or other items with solar energy and magnifying lenses. Such is the versatility of concentration. Typically, a tube is placed in the focal area, with a liquid circulating inside to carry the heat away. Concentrators are required when temperatures higher than about 400 K are sought. The efficiency of this type of collector seems to lie in the 50 to 75% range.

The most promising application of concentrating collectors (from a thermodynamic viewpoint, at least) is that of power generation. The concept relies on generating a vapor of sufficient enthalpy to produce work when expanded in a turbine. Other prime movers, such as the venerable Stirling engine running on air, have been tested with concentrating collectors, but the qualities of turbomachinery in terms of energy density seem to be preferred for concept implementation. In principle, the necessary

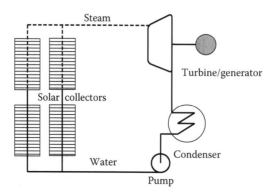

FIGURE 5.20 Concept of solar thermal power plant.

area of solar collectors is deployed, with concentrating capabilities to heat the fluid (typically steam) to 800 K at pressures on the order of 100 bars (Figure 5.20). These are stringent conditions, but the qualities of solar energy enable access to those temperatures. The area requirements tend to be large depending on the average insolation, as shown by Example 5.7.

EXAMPLE 5.7 THERMAL AREA REQUIREMENT

For the average insolation of 5 kW · hr/(m² · day) adopted in this chapter, estimate the area required by a modern 100 MW plant with efficiency of 0.35. Use the average insolation, although keep in mind that, to capture the solar peak values, the area required could be larger.

Solution

A 100 MWe plant operating at 0.35 efficiency will require a thermal input (*TI*) of

$$P = 100 \text{ MWe} \qquad \eta_{th} = 0.35$$

$$TI = \frac{P}{\eta_m} = \frac{100}{0.35} = 285 \text{ MW}$$

Now, the insolation of $PVW_{USA,}$ is

$$PVW_{USA} = 5 \cdot \frac{\text{kW} \cdot \text{hr}}{\text{day} \cdot \text{m}^2} = 5 \cdot \frac{1000 \cdot \text{W} \cdot \text{hr}}{24 \cdot \text{hr} \cdot \text{m}^2} = 208 \ \frac{\text{W}}{\text{m}^2}$$

Hence, the area required would be

$$A = \frac{TI}{PVW_{USA}} = \frac{285 \cdot \text{MW}}{208 \cdot \dfrac{\text{W}}{\text{m}^2}} = 1.37 \cdot 10^6 \text{ m}^2$$

This collector area is 0.53 of a square mile (a square of 0.73 mi side), which is not a large area. The electric output per unit area is given by

$$SPO = PVW_{USA} \cdot \eta_{th} = 208 \cdot \frac{W}{m^2} \cdot 0.35 \cdot \frac{We}{W} = 72.8 \frac{We}{m^2}$$

The following two examples (Examples 5.8 and 5.9), illustrate the dynamic nature of solar energy. As the Sun moves across the sky, the transient nature of the arriving energy must be dealt with for effective capture.

Example 5.8 The Sun in Motion [VisSim]

With the respect due to the most venerable of all energy sources (heavy isotopes might have predated it, but they were not used until 60 years ago), and the only one capable of nuclear fusion, we study in this example the trajectory of the Sun and calculate the angle that its rays form with a flat surface of given orientation.

The surface is fixed, and being flat, its position can be defined via two angles (Figure Example 5.8.1): μ is the angle that the normal to the surface makes with the vertical of the location. The second angle, ψ, is the angle formed by the N–S line in a horizontal plane and the projection of the normal in the same plane.

With the surface thus defined, the position of the Sun relative to the normal will vary with time (along the year with each month, with each day, and within the day, each minute). We seek to identify the direction of the Sun's rays via the angle that they make with the normal. The angle is labeled θ, and it is shown in Figure Example 5.8.2, along with a new angle, Φ, which is the angle formed by the projection of the Sun's ray direction onto the horizontal plane and the N–S line. Our ultimate goal is to calculate the radiation received by the flat surface, and this requires knowledge of the θ angle at every instant of daylight.

The first input to the VisSim program for this example is the latitude. Other inputs concerning the location are the day of the year (a summer day was chosen) and the

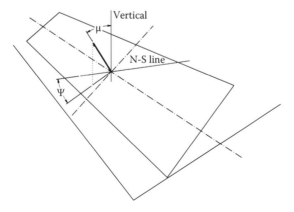

FIGURE EXAMPLE 5.8.1 Angular coordinates defining the position of a surface.

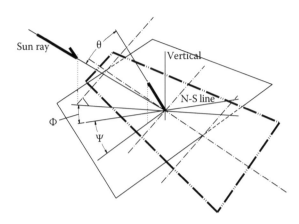

FIGURE EXAMPLE 5.8.2 The angle θ between the normal to the surface and the Sun's rays. The collector perimeter is depicted by dash-dot-dot line.

duration of sunlight corresponding to the chosen latitude and day [1]:

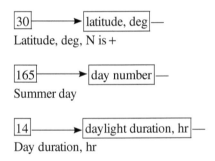

As discussed, two angles define the orientation of the collector, and the following was chosen:

Collector orientation

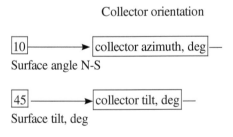

With the inputs completed, we proceed now to calculate the declination δ, namely, the angle formed (like all angles) by two lines in a plane. The first line (Figure Example 5.8.3) joins the center of the Earth (E) and of the Sun (S). The plane is defined by the first line and the axis of the Earth. This plane intersects the equator at point Q. The second line joins E and Q.

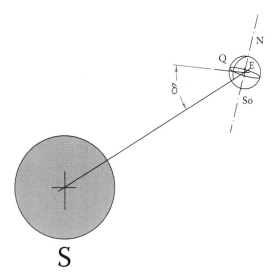

FIGURE EXAMPLE 5.8.3 The angle δ measures the elevation of the Sun as the Earth travels its orbit.

The declination is given by Reference 2:

$$\delta = 23.45 \cdot \sin\left(360 \cdot \frac{284 + n}{365}\right) \tag{a}$$

and programmed in the block

$$\longrightarrow \boxed{\text{Declination for given day}}\!\!-$$

The solar hour angle H can be visualized via projection of two lines on the equatorial plane via meridians. One line is SE, already found in Figure Example 5.8.3 and now also in Figure Example 5.8.4. The other is the normal to the location under study,

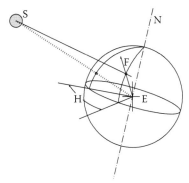

FIGURE EXAMPLE 5.8.4 The hour angle H is defined with respect to the local meridian.

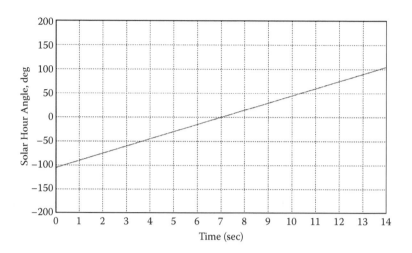

FIGURE EXAMPLE 5.8.5 Variation of solar angle versus time.

labeled F in honor of the Flatlanders [3]. The projection of both lines onto the equator yields two lines with the solar hour angle H in between.

The angle H is negative before the solar noon, which occurs when H is zero. In the afternoon the angle is positive. The block

$$\longrightarrow \boxed{\text{Solar angle for given duration}}\!\!-$$

performs the rather simple function of dividing the daylight interval into two equal ones and translating the hours thus obtained into degrees since each hour corresponds to 15° at the rate of rotation of the Earth. As the day is simulated in VisSim, the simulation time starts at zero and covers the daylight time in hours. The solar clock starts at sunrise (or shortly before it to account for dawn), and progresses to zero (at solar noon) to finish after dusk (Figure Example 5.8.5).

The solar altitude angle, β, measures the altitude of the Sun over the horizon. We show in Figure Example 5.8.6 by dash-dot line the horizon plane on which the inhabitants of the legendary Flatland perceive their lives developing. The line F–E is the vertical to the horizon plane, and the vertical projection of the Sun onto this plane defines the point P. The angle between FP and FS_0 (North–South line in Flatland) is Φ, the solar azimuth, already found in Figure Example 5.8.2. In a plane perpendicular to the Flatland plane, namely the plane SPF, we define two angles: one is the solar altitude β, and the other is its complementary, α. The angle β is given by Reference 2:

$$\sin(\beta) = \cos(l) \cdot \cos(H) \cdot \cos(\delta) + \sin(l) \cdot \sin(\delta) \tag{b}$$

where l is the latitude, H is the hour angle, and δ is the declination. This angle is computed in the block

$$\boxed{\text{Compute altitude angle Beta}}\!\!-$$

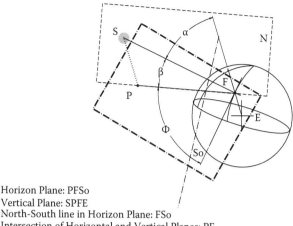

Horizon Plane: PFSo
Vertical Plane: SPFE
North-South line in Horizon Plane: FSo
Intersection of Horizontal and Vertical Planes: PF
β: Angle in Vertical Plane, between SF and PF
Φ: Angle in Horizon Plane, between FSo and PF

FIGURE EXAMPLE 5.8.6 The angle β measures the altitude of the Sun over the local horizon.

The azimuth angle is also easily computed as [2]

$$\cos(\Phi) = \frac{\sin(l) \cdot \cos(H) \cdot \cos(\delta) - \cos(l) \cdot \sin(\delta)}{\cos(\beta)} \quad \text{(c)}$$

This equation is implemented in the block

Compute solar azimuth Phi

We now calculate the following angle [2]:

$$\gamma = \left| \Phi - \Psi \right| \quad \text{(d)}$$

where Φ is defined in Figure Example 5.8.6, and Ψ in Figure Example 5.8.1. Finally, we arrive at the object of our labors, namely θ, the angle between the Sun's rays and the normal to the surface [2]:

$$\cos(\theta) = \cos(\beta) \cdot \cos(\gamma) \cdot \sin(\mu) + \sin(\beta) \cdot \cos(\mu) \quad \text{(e)}$$

where μ is defined in Figure Example 5.8.1. As daylight progresses, we can then track the θ angle in Figure Example 5.8.7.

As shown in the next example, knowledge of this angle is necessary to calculate the direct solar radiation incident upon the surface.

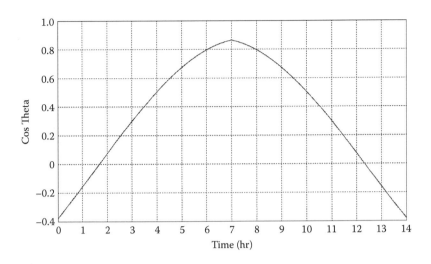

FIGURE EXAMPLE 5.8.7 Variation of (*cos θ*) versus time.

REFERENCES FOR EXAMPLE 5.8

1. http://encarta.msn.com/media_701500905/Hours_of_Daylight_by_Latitude.html
2. Kuehn, T. H., Ramsey, J., Threlkeld, J. L., 1998, *Thermal Environmental Engineering*, Prentice Hall, NJ.
3. Abbott, E., 2004, *Flatland: A Romance of Many Dimensions*, Barnes & Noble, NY.

EXAMPLE 5.9 CAPTURING THE ENERGY OF THE SUN IN MOTION [VISSIM]

Learning about the possibilities of solar collectors implies factoring in transient analysis of the Sun–collector system.

Our Sun (a fusion reactor contained by its own gravitational pull) emits energy at an incredible rate: $4 \cdot 10^{23}$ kW [1] all of the time, and the flux at the Sun's surface is 63 MW/m². For comparison, a 1-GW coal-fired plant may require 20 acres for the generation equipment for a flux of 12 kW/m². The energy flux from the Sun decreases with the square of the distance from the Sun (Figure Example 5.9.1). By the time we get to the Sun–Earth distance R, the flux (or irradiance) is 1.37 kW/ m². This flux is what a satellite experiences. The radiation now travels through the atmosphere, where it is scattered and absorbed in various measures by different molecules and matter (Figure 5.16). It can also be reflected by clouds. The flux of energy arriving directly from the Sun is called *direct irradiance*.

So, on a cloudless day, a surface oriented towards the Sun, at noon, could experience direct irradiance of about 1 kW/m² or 25% less than the satellite in outer space. Part of the scattered energy can also arrive at the surface as diffuse radiation. The sum of direct irradiance and diffuse radiation is called the *global irradiance*. Of course, this definition assumes optimal conditions. Under real conditions, the average amount is much less than the global irradiance due to clouds and other obstructions (such as dust or shade over the collector), which must be duly factored in for a suitable design.

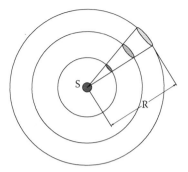

FIGURE EXAMPLE 5.9.1 As radiative energy emitted by the Sun travels in space, the energy flux decreases with the square of the radius *R*.

We study in this example the energy collected and the transients of a partially cloudy day experienced with a solar collector.

The global irradiance is then the sum of the direct irradiance multiplied by the *cos* (θ) (laboriously acquired in Example 5.8) and the diffuse radiation multiplied by the fraction that actually strikes the collector. For computing the irradiation and diffuse radiation, we use the model advanced in Reference 1. Whereas the *cos* (θ) was calculated in Example 5.8, we add to the inputs of that example four new variables, namely the area of a collector, and various coefficients to calculate the irradiation and diffuse radiation.

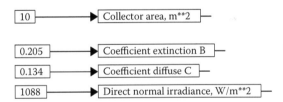

The direct irradiance *DI* is given by

$$DI = \frac{A}{\exp\left(\dfrac{B}{\sin(\beta)}\right)} \tag{a}$$

where *A* is the direct normal irradiance, *B* is the extinction coefficient, and β is the solar elevation angle of Example 5.8:

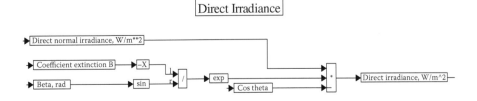

The irradiance is multiplied by the cos (θ) to define the direct irradiance incident upon the collector. The block

Diffuse Solar Radiation

computes the diffuse radiation DR on a horizontal surface as a simple relationship

$$DR = C \cdot DI \tag{b}$$

where C is the coefficient of diffuse radiation of the adopted model [1]. The radiation actually arriving to the collector is the diffuse component DC, given by

$$DC = DR \cdot \frac{1 + \cos(\mu)}{2} \tag{c}$$

where μ is the angle between the vertical and the normal to the surface, defined in Example 5.8. The global irradiance (i.e., the sum of direct irradiance and diffuse radiation) is computed in block

Global Irradiance

To illustrate the problems of storage and compatibility with the network, we compare the energy arriving at the collector on a clear and on a semi-cloudy day. (In a completely overcast day, only the diffuse radiation would count.) The simulation of a semi-cloudy day may take various forms. We chose here to simulate the day by generating random pulses of random duration with the block PRBS. At each rising pulse, a Sample-Hold block records the value of a Gaussian signal of mean value equal to 0.5 and standard deviation of 0.05. The Gaussian signal is multiplied by half of the PRSB pulse, and the result (magnified by 3) is subtracted from the sum of half of the PRSB pulse and 1. The gains half and 3 have been calibrated to yield a signal between 0 and 1. The blocks employed for this purpose are shown below:

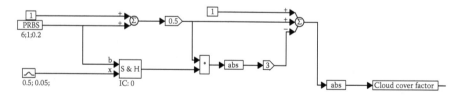

The result of these operations is the cloud cover factor, a random signal that obscures 50% of the global irradiation when a cloud is present (Figure Example 5.9.2) and depicts periods of cloudless sky and clusters of passing clouds. On a perfectly clear day, the cloud cover factor is set to one.

The product of the global irradiance, cloud cover factor, and collector area gives the net power arriving at the collector. The integration of such power yields the

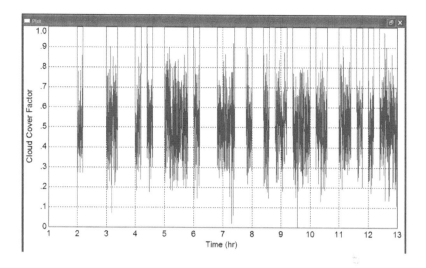

FIGURE EXAMPLE 5.9.2 As clouds pass over the collector, solar energy is deducted to account for clouds.

energy arriving at the collector. These computations for the semi-cloudy and clear day are carried out in the following block:

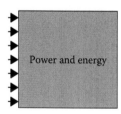

The ratio of energy received on the assumed semi-cloudy day to that on the clear day is shown below:

The times and reduction of available power are of interest for grid and backup system design.

REFERENCES FOR EXAMPLE 5.9

1. Kuehn, T. H., Ramsey, J., Threlkeld, J. L., 1998, *Thermal Environmental Engineering*. Prentice Hall, NJ.

BIOMASS

Biomass is a means of capturing solar energy. Different ways to use the product exist; we describe the main ones here. Because of its importance to the human endeavor, we highlight here the technology for biomass production and combustion. Liquid fuels can be and are being produced from biomass. Because the energy yield of liquid fuel production is unclear (except for ethanol from sugarcane), we refer only briefly to this application.

Wood can be burned as chips or sawdust. Woodchips vary in size, with typical dimensions ranging from 6 to 14 mm. They are typically burned in boilers not unlike the ones used for coal, with a moving grate burner (Figure 5.21). The wood chips flow from a hopper to a moving grate where combustion occurs. The grate openings are such that ash is collected at the end or falls between the grates. Not unlike coal burning, wood burning requires provisions for ash removal. The steam can be used for heating or power production. Drying the wood is most helpful for good combustion and high heating values. A moisture level not exceeding 15% by weight is a common target.

Sawdust combustion typically requires a different burner, of torsional chamber design, suitable for other fuels as well [6]. Typical configurations (also used for coal) inject the fuel and air tangentially in a cylindrical chamber with water-cooled walls (Figure 5.22(a)). Sustained by correct air velocities, the particles follow helicoidal trajectories until completely burnt. Ash is collected in the chamber's walls and also in the bottom of the boiler plenum. The plenum of the boiler can have waterwalls, or the gases can be routed to flow among tubes (Figure 5.22(b)). Fuels other than wood [6,7] are used in these boilers, but wood is a common source of renewable energy for these designs. The most widespread use of wood energy is via liquid fuels. The Kraft paper-making process results in a liquid containing most of the lignin originally present in the wood feedstock. This liquid, called black liquor, is burned in boilers to produce steam for process or power applications.

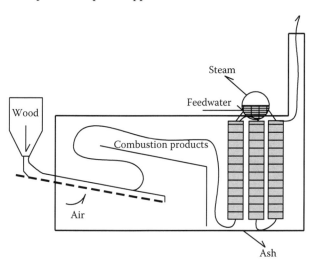

FIGURE 5.21 A moving grate boiler.

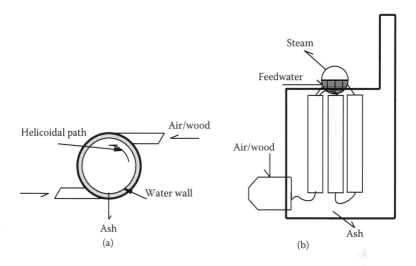

FIGURE 5.22 Torsional chamber concept (a) shows the chamber, (b) chamber position in the boiler.

An important, emerging application of biomass is the production of liquid fuels, of paramount importance for transportation. Corn is used to produce ethanol by fermentation of its sugars. The output is a mixture of solids, water, and ethanol (about 7%).

The alcohol is concentrated via distillation, and the remaining water can be extracted when blended with gasoline, for ethanol mixes with gasoline but water does not and it is readily separated. The 10–90% proportion (alcohol–gasoline), results in a liquid fuel that all engines can tolerate. More alcohol requires engine modifications, but this could very well be an emerging trend. Substantial information is offered in Reference 8.

Of the U.S. energy budget, namely, 105 EJ, biomass supplied 3 EJ in 2004. One could wonder why this text describes such a pittance. The reason is oxygen. For eons, biomass has engaged in the following reaction to grow:

$$6H_2O + 6CO_2 + sunlight \rightarrow C_6H_{12}O_6 + 6O_2$$

that is, biomass captures carbon dioxide and water, storing energy as glucose and releasing oxygen. This oxygen is necessary for almost all life forms. Photosynthesis is the name of this reaction; mankind did not invent it, it is responsible for the low levels of CO_2 in the atmosphere, and it has proven impossible to date to reproduce it in useful scale to mitigate our pollution problems. It is, beyond doubt, the most significant equation of this book, and the process it represents is crucial for this world.

HYDRO

We include hydro here because it is the quintessential cycle. It could not occur without solar energy and gravity, but its unique blend of thermal and mechanical aspects predates all other cycles. Thermal energy from the Sun heats up bodies of water,

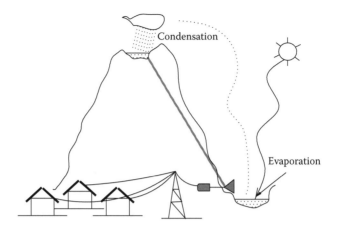

FIGURE 5.23 The basic hydro cycle.

evaporating the water on the surface. The water vapor flows upwards until it condenses due to the lower temperatures above the Earth's surface. Eventually, the liquid water returns to Earth as rain. It is the job of engineers to capture this water in reservoirs and pipe it to turbines at a lower level than the water free surface (Figure 5.23). It is clear that, like most thermal cycles, hydro requires a hot and a cold temperature to work, the former to evaporate the water, the latter to condense it and to form clouds.

The key machine is a water turbine, of which there are several types for different operational envelopes. For high heads and low flows, impulse turbines are preferred. *Impulse* means that the change in momentum of the flow is primarily responsible for power production. For instance, we illustrate the process in Figure 5.24(a). *Reaction* takes place as a consequence of the flow accelerating between the blades of the turbine (Figure 5.24(b)). The typical impulse turbine is the Pelton wheel, of long fame in the Wild West movies depicting the gold rush. A technology vastly in use today is the Kaplan turbine, shown as a reaction type turbine.

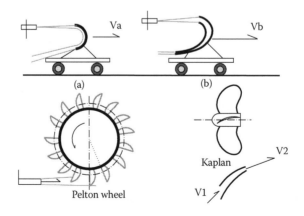

FIGURE 5.24 Hydro turbine concepts (a) impulse (b) reaction.

As opposed to biomass, hydro does not capture CO_2 or create O_2, but it does save fuel and it eliminates CO_2 or nuclear waste that would have been associated with an equivalent $KW \cdot hr$ produced.

AREA: NOT A SUPERFICIAL TOPIC

Renewables do not consume fuel but take up surface area. Real estate is a finite commodity like fossil and nuclear fuels, but it is much easier to see how it is finite. So, there is some degree of competition for the area required. For instance, photovoltaic and solar thermal may use rooftops unused as of now. Hence, there is no infringement of area usage. Large windmill farms clearly cannot be installed in urban areas, although some farm activities are compatible with windmills. Their preferred use probably lies in coastal areas, on water surface otherwise unused. Biomass competes head-on with production of food. When what is burned is refuse from food production, there is no conflict. However, there is no doubt as to the preferred use when it boils down to avoiding hunger. Finally, hydro takes up some serious area. The dams produce lakes that can be used for recreation and perhaps commercial fishing, but other than irrigation head, there is little one can do with an artificial lake other than enjoy, and the latter may make all the difference for a well-lived life.

SUMMARY

From advanced materials with carefully crafted conduction bands to boilers for biomass, passing through advanced coatings, specialty glass, and novel airfoils, renewables have made a lot of progress. Renewable technologies cope with the low energy density of the resource in any of its forms: solar, wind, or biomass. Some may add that they cannot cope with the low density of money in many households that clearly could not afford PV. Thermal may be affordable, and many see wind as economically viable. Biomass has been and still is a fuel for many industries and residences throughout the world.

Renewable technologies have to cope with another matter: many resent its "low-tech" image and its undeniably whimsical nature, which storage can only partially erase. We can only offer a vision. Humans have gotten this far because of remarkable ingenuity, which was not given to us yesterday. The grit to survive and the determination to face the future with all our intellectual and physical resources are our trademark. Renewables are an opportunity to use our ingenuity for a better world. It is certain that their widespread use will require adjustments, but should those adjustments occur, we will perhaps come to realize that this future, of all the alternative ones, was certainly the best.

All in all, PV is certainly an advanced technology of great potential, whereas solar thermal is an easy, reliable way to save fossil energy. Wind has been growing considerably and may hold a substantial promise. Biomass is in the spotlight: its conversion to biofuels is expected to reduce oil dependence even if that happens through the agency of coal or nuclear. Biomass combustion is an old reliable energy transformation technology, perhaps not receiving as much attention as it should.

REFERENCES

1. Hayden, H. C., 2001, *The Solar Fraud*, Vales Lake Publishing.
2. EIA, Energy Information Agency, 2008, http://www.eia.doe.gov/cneaf/solar.renewables/page/solarphotv/photovoltaics2.gif
3. NREL, National Renewable Energy Laboratory, 2008, http://www.nrel.gov/gis/images/us_pv_annual_may2004.jpg
4. Danish Wind Industry Association, http://www.windpower.org/en/tour/design/concepts.htm
5. Center for Space Research, http://www.csr.utexas.edu/projects/rs/hrs/pics/irradiance.gif
6. Agrest J., 1973, Firing Chamber for the Combustion of Gaseous, Liquid or Fine-Granular Fuel. United StatesPatent # 3718122.
7. Tejero, I., Trujillo, A., León, E., 2000, An Efficient Technology for The Combustion of Biomass, 1st World Conference and Exhibition on Biomass for Energy and Industry, 5–9 June, Sevilla, España.
8. Luhnow, D., and G. Samor 2006. *As Brazil Fills up on Ethanol, it Weans Off Energy Imports,* the Wall Street Journal, Jan. 16, 2006.

6 Après Conversion
Utilization Technology

Ah yes! Ask the infinity to come in!

Louis Aragon in *Libro de Sueños*, **J. L. Borges**

All the production and conversion processes feed end users, a definition that encompasses most of us. The energy consumed by each sector (residential, commercial, industrial, and transportation) is presented. Much of the electric power goes to drive electric motors, and we cover their operational principles. Lighting is addressed here as well, as are refrigeration/heating cycles and heating devices. Commercial uses of microturbines and fuel cells are explained. The processes whereby steel, aluminum, paper, and plastics are obtained are summarized. The dynamic simulation of a small building, showing the influence of insulation, equipment efficiency, and thermal mass is presented.

There are four broad sectors to an economy: residential, commercial, transportation, and industrial. We describe generically important technologies that each sector uses, with an eye towards determining which ones, if any, can make a substantial difference in the energy future. We use here the more customary SI units for energy, and we give equivalences to the units of Chapter 3: 1 EJ = 0.024 btoe; 1 MJ = $2.38 \cdot 10^{-5}$ toe.

RESIDENTIAL

According to Reference 1, the residential energy use follows the patterns of Figure 6.1 in the United States. We show there the total energy, and the energy used for heating, cooling, and hot water. A number of appliances and machines are not reflected in this use, and the total for electricity is above the sum of the three uses noted. This is not the case for natural gas: since there are no PCs or lights or tools directly activated by natural gas, its use is confined to heating and hot water.

The residential needs encompass heating, cooling, laundry, food conservation and preparation, lighting, information technology, and small tools. The common denominator to all of these needs is electric motors, which we describe here. Refrigeration and heating cycles (heat pumps and air conditioners) come next. Gas/oil furnaces, woodstoves, and some reference to lighting complete this description. Dynamic modeling is useful here to show how transients dictate the controls and periods of usage of technology.

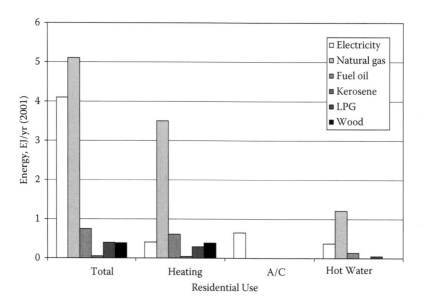

FIGURE 6.1 Residential energy use.

Electric Motors

An electric motor is, like a fuel cell, a direct energy-conversion device, converting from electrical to mechanical energy. (A fuel cell converts from hydrogen to electrical.) The basic principle of the electric motor relies on the observed phenomena that magnetic dipoles align themselves with the external magnetic field. This was discovered in China in about 250 B.C., and about one thousand years later, magnets were used as guidance aids in ships. The basic idea is illustrated in Figure 6.2. If a magnetic needle forms an angle α with an external magnetic field, a torque is exerted on the needle until the angle α is zero and the north (N) of the needle points to the south (S) of the external field.

The trick is in how to use this principle continuously, for which one would need the external field to rotate so as to maintain the angle α constantly. This can be done with AC and judiciously placed coils that generate a rotating magnetic field. Consider Figure 6.3, which shows a coil subject to AC voltage. The coil creates a magnetic field in the magnetic circuit shown. As the current alternates in direction, so does the direction of the field in the circuit. Imagine now that a magnet is inserted in the rotor as shown. At the instant that a north pole (N) exists at the bottom, the magnet will rotate to align its N with the corresponding south (S) in the magnetic circuit called the stator. If as the magnet is arriving, the current is inverted and a N forms at the top of the stator, the rotor will continue turning towards the bottom pole. If the rotor is connected to a load, such as a washer or dryer or refrigerator compressor, it will deliver work to the load. Hence, the basic principle consists of generating two magnetic fields: one of which moves in space and the other of which follows it, with the consequent generation of torque. The virtual technical universe abounds with machines that will run, if they could only start. In the motor just described, the

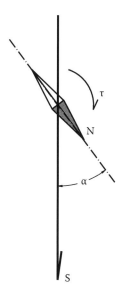

FIGURE 6.2 Compass.

direction of rotation could be any of the two possible ones upon starting, depending on the position of the rotor. Auxiliary coils, typically in series with a capacitor, are added so that the stator magnetic field builds up to result in the preferred direction of rotation. The capacitor is taken out of the circuit by a switch when the motor picks up speed. In addition to starting transients, other problems arise with the concept just described, notably concerning the generation of the necessary magnetic fields. Rotors with permanent magnets such as the one used in our description are becoming more common, but they would be too expensive and would not offer, in some cases, enough torque. The solution is surprisingly simple: to use a squirrel cage rotor (Figure 6.4). In this rotor, there are closed coils that interact with the stator magnetic field.

Lenz's law establishes that a varying magnetic field through a coil will induce a voltage such that the resultant currents will tend to oppose the decay or growth of the magnetic field. The rotating coil of the rotor is subjected to the external magnetic

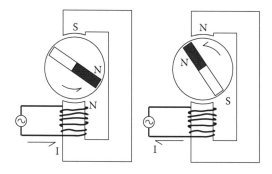

FIGURE 6.3 Single-phase motor.

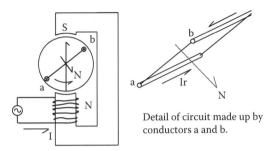

Detail of circuit made up by conductors a and b.

FIGURE 6.4 Principle of squirrel cage motor.

field, produced by the stator (Figure 6.4). When the current is increasing, the stator field is increasing, and, by Lenz's law, the current in the coils will create a field to oppose this increase. This is happening just when the coil is moving toward the stator poles. Upon entering the strong field close to the poles, there is a force acting on the conductor carrying the current Ir, as a consequence of the stator field. The force (which is the vector product of the current multiplied by the stator field) keeps conductor (a) rotating in the counterclockwise direction. The same thing happens at the same instance to conductor (b). Thus, once started, the motor keeps on rotating, and it also can support a torque. However, for the torque to arise, the rotor has to lag behind the stator magnetic field so that the current Ir will peak in the region with the stronger field. The torque is roughly proportional to the lag in speed. This type of motor is therefore called asynchronic. Clearly, the AC frequency creating the stator field will control the rotational speeds. An inverter can be used to control the frequency and hence the speed of these motors.

Another common type of motor in households is the universal type, so called because it can work on DC or AC current. In the universal motor, the rotor magnetic field is built by connecting a coil to a commutator and brushes (Figure 6.5). The commutator is composed of a number (two in the figure) of cylindrical sectors

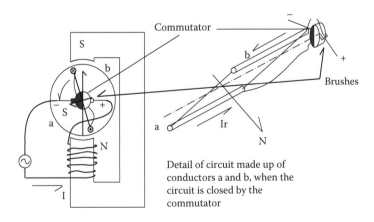

Detail of circuit made up of conductors a and b, when the circuit is closed by the commutator

FIGURE 6.5 Universal motor principle.

isolated from each other. As the commutator and rotor rotate, the cylindrical sectors corresponding to one rotor coil are connected to the external voltage. A current flows from positive to negative through the coil, producing a rotor magnetic field. The torque acting on the coil is such as to align the rotor S and the stator N. The brushes rub against the commutator, allowing current to flow to the rotor coil. When the rotor has completed another 180°, the rotor field is still oriented as shown because the current still flows in the same direction, although in different branches on the rotor coil. Commutators have many circular sectors, and hence, the spatial variation of the rotor field is minimal. Since the rotor and stator are connected in series to the line voltage, simple voltage variation changes both the stator and rotor fields. The rotational velocity depends on the field strength, and hence, the voltage control is an easy way to control the motor speed. The voltage is typically controlled with variable resistors, as in the triggers of many of these motors in which the trigger position results in different resistor values. (Other, more efficient controls are common, for resistors are great ways to generate entropy or increased utility bills.)

Other motors, notably those in PCs, consume a fair amount of power. We do not describe these DC motors in detail here. Suffice it to say that the rotor has permanent magnets, and that the stator has, typically, three coils. Electronic controls feed current to the coils in sequence so as to create a truly rotating field, which the rotor magnets follow, thus resulting in rotor motion. These motors are brushless and tend to have higher efficiencies than universal motors. Improving motor efficiency could go a long way toward energy savings.

Synchronous motors were once of incredible importance in households, for they activated electric clocks. Before the advent of electronics, this was an important function, for winding up clocks (another venerable energy conversion/storage function) has never been a favorite of mankind. A brief history of their significance, both technical and social, can be found in Reference 2. The principle of synchronicity might be apprehended by following Figure 6.6. Imagine three coils like the stationary ones

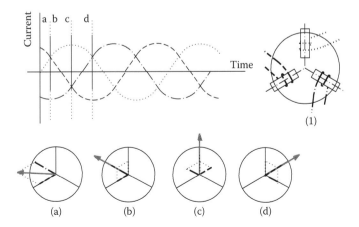

FIGURE 6.6 A rotating magnetic field arising from three stationary magnetic fields, spaced at 120° in space and time.

in Figures 6.5 or 6.4, placed pointing at the center of a circle and spaced 120° (inset 1 in Figure 6.6). Each coil is activated by current (dot, dash or dash-dot). The currents do not increase or decrease simultaneously, but each has its own phase, separated by 120° from the other two. Thus, each coil has its own magnetic field that changes periodically following the current sinusoid, each 120° out of phase with the others. The consequence is three magnetic fields (dot, dash or dash-dot) that add up to the rotating vector, which is then the magnetic field of the stator. As time flows from (a) to (b), the fields evolve in such a way that the magnitude of the rotating vector is constant but the direction has changed. Further, from (b) to (c) and then (d), we observe that the magnetic field has rotated clockwise as time has flowed, as it habitually does, forward. The rotating magnetic field then drives the rotor, which has a magnetic field produced either by a permanent magnet, or, in case of larger motors, by running current through a coil fastened to the rotor. The coil receives direct current via slip rings, and hence, the rotor field is stationary with respect to the rotor. Both magnetic fields (stator and rotor) line up, and the rotor rotational speed coincides with the rotating speed of the stator magnetic field.

Unlike an asynchronic motor, a synchronous motor has a rotor that stays in perfect step with the field unless the torque exceeds the motor capabilities, in which case the motor ceases to operate normally. Today, synchronous motors have wide-ranging applications in industrial settings, where speed control or controlling the current drawn from the net is necessary.

Heating/Cooling with Electricity

A variety of devices are used to provide space conditioning. For heating, resistors (of poor efficiency; see Chapter 7), are used in some applications. Whereas air conditioners have been used for a long time, heat pumps have been developed more recently, and are increasingly applied. We describe a simple refrigeration cycle (Figure 6.7) consisting of an evaporator, condenser, compressor, and expansion valve, a cycle that is used by heat pumps and air conditioners. Electric motors, described earlier, fill a crucial role in these technologies: there are two fans in most heat pumps, each with its electric motor, as well as an electric motor that activates the compressor. Increasingly, all motors are of variable speed to follow varying loads.

From 1 to 2, the refrigerant evaporates with heat addition. From 2 to 3, the refrigerant vapor is compressed to result in superheated vapor discharge. From 3 to 4, the vapor is condensed into liquid. The liquid refrigerant flows to a reservoir (accumulator). The reservoir communicates to the evaporator via an expansion valve. The opening in the valve responds to the capacity desired in the evaporator. As the capacity increases, control strategies open the valve more, allowing further flow of refrigerant. The state points are illustrated in Figure 6.7(b), in a TS diagram.

The point is that the refrigerant receives heat in the evaporator. Hence, in winter, if it is desired to harvest heat from the outdoors, the fan blows outdoor air over the evaporator coil, with the refrigerant extracting heat from the air. The heat absorbed by the refrigerant, plus that corresponding to the enthalpy change due to the compressor work, is delivered to the house at a higher temperature (points 3 to 4).

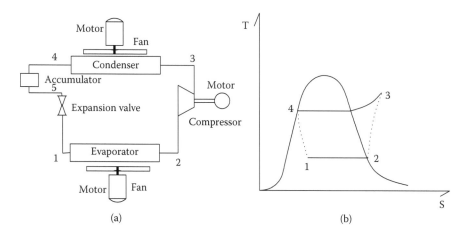

FIGURE 6.7 Cooling/Heating cycle, (a) components, (b) state points in TS diagram.

The ratio of the thermal energy absorbed to that delivered is called the Coefficient of Performance, or COP.

COP values can be low when the outdoor temperature is low, but they increase as the outdoor temperature increases. For air-source heat pumps, COPs range from 1 (i.e., resistance heating) to 4 or even 5. The synergy between a little high-grade energy to upgrade the temperature of a lot of low-grade energy (up to five units) is remarkable.

The same cycle is used for cooling. In the summer, the evaporator is inside the house. As the refrigerant evaporates, the house is cooled. The heat is rejected outdoors in the condenser. Because the temperature differences between outdoors and indoors are more limited in the summer than in some winter locations, the summer COPs tend to be high, from 3 to 5. If you are wondering why you only see one machine per house, it is because ingenious valving changes the circulation of the refrigerant in such a way that the same heat exchanger acts as the evaporator (i.e., outdoor heat pump) in the winter, and condenser (i.e., outdoor A/C) in the summer.

Other than compressor and motor efficiencies, heat pumps and HVAC must meet varying loads. As time goes by, weather and occupancy change, and so does the load. Controlling machines has become a crucial aspect of energy management. Advanced units now on the market control compressor and fan speeds to follow the load continuously, as it will be shown in the thermostat setback example at the end of the chapter.

LIGHTING

Lighting is fundamental to a healthy intellectual endeavor. Incandescent lighting has low efficiencies. The theory behind this lighting is to heat a tungsten filament to 2300 K or so, at which point the tungsten is so hot that it radiates in the visible spectrum. Of the power given to a bulb, no more than 6% results in visible radiation. We offer other efficiency measurements later on. In any case, the tungsten filament is either in vacuum or in a mixture of argon and nitrogen, which slow down the evaporation of the filament.

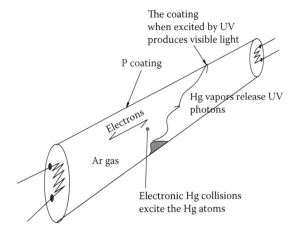

The coating
when excited by UV
produces visible light

P coating

Hg vapors release UV
photons

Electrons

Ar gas

Electronic Hg collisions
excite the Hg atoms

FIGURE 6.8 Schematic of a fluorescent lamp.

Fluorescent lighting has a larger initial cost, but it saves energy and it is more efficient than incandescent lighting. In fluorescent lights, electronic excitation is responsible for light emission, and the emitters (mercury [Hg] and phosphorus) are not heated, as tungsten is in incandescent bulbs. Hence, the process is more efficient. We illustrate the principle in Figure 6.8. Electrons from the electrode are accelerated by a voltage difference between electrodes, in a mixture of inert gas and Hg vapor. Electronic collisions excite the Hg vapor atoms. (Hg vaporizes easily.) Upon returning to their nonexcited state, these atoms emit UV light. The UV light impacts the P atoms on the coating. These atoms are then excited, and upon returning to the ground state, they emit in the visible range. Initially, a large voltage is required (500 to 700 V), but as the electronic current is established, the required voltage decreases.

Fluorescents have much larger efficiencies than incandescents. Consumers are interested in how light fills their space and how they perceive it to do so. Hence, the radiative power of a light source is codified with reference to the human eye by a rather complicated equation [3], which gives the total radiative power in lumens. In a way, lumens is somewhat like a "watt," for it measures power and spatial distribution as perceived by a standard eye. In any case, the efficiency is then in lumens/watt, the lumens being the light output and the watt being the electrical power input. For incandescents, this value is around 30 lumens/watt, whereas it is up to 100 lumens/watt for fluorescent lights.

Many anticipate that light-emitting diode (LED) technology will soon make serious market inroads. In Figure 5.4 (Chapter 5) we showed how sunlight creates excess electrons in the n-type crystal of a diode. The p–n junction was essentially saturated, and electrons found it easier to flow through the load to the p-type, than to flow directly across the junction. In effect, sunlight is converted into electric power. Someone [4] presumably wondered if the inverse would be true, that is, if application of a voltage to a diode would make it produce light. This turns out to be the case for gallium arsenic (GaAs) or gallium nitrogen (GaN) crystals.

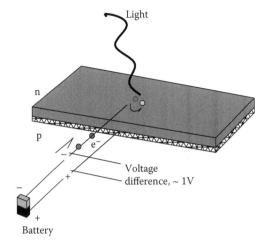

FIGURE 6.9 LED principle.

When a GaAs crystal is doped with Zn, it becomes a p-type semiconductor, and doped with selenium (Se), it becomes an n-type conductor. When the two types are placed adjoining each other, a junction region where the electrons fill the holes arises. Overcoming this region requires an external voltage, as shown in Figure 6.9, greater than 1V. The electrons in the n region now can overcome the junction region. (They have enough energy, as given by the battery, to do so.) Upon flowing to the n junction, they recombine filling up holes. This recombination emits light, the frequency of which depends on the crystal and doping. For concentrated, spot-type lights, LEDs are efficient. For illuminating broad areas, the efficiency gains do not appear to be that clear [5].

LEDs have found applications in the automotive industry for lights in cars. Also, traffic lights, requiring a low-maintenance, concentrated beam, have adopted LEDs. Perhaps their reach will broaden in the future, especially if a vibrant solar photovoltaic industry develops.

HEATING WITH NATURAL GAS

The composition of natural gas varies somewhat, but the essential combustion reaction is (Chapter 4)

$$CH_4 + 2 \cdot (1+e) \cdot (O_2 + 3.76\,N_2) \rightarrow 2\,H_2O + CO_2 + (7.52 + 2 \cdot e)\,N_2 + 2 \cdot e \cdot O_2$$

where there is the possibility of excess air, e. The point is that natural gas produces two moles of water per each mole of methane. Water is hydrogen already oxidized; one cannot expect much energy from an oxidized substance. The water, however, is in the vapor phase, and when water vapor turns into liquid, a great deal of energy is released. The heat of condensation of water at low pressures is on the order of 2500 kJ/kg, a rather respectable value compared to those of all other refrigerants. Recovering the latent heat contained in the water vapor is worthwhile.

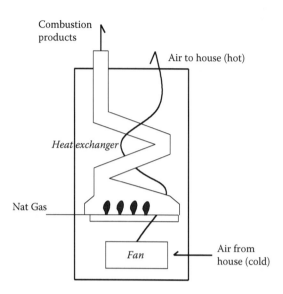

FIGURE 6.10 Gas furnace, noncondensing.

Hence, it is necessary at this point to define two technologies: noncondensing and condensing ones. The latter have larger efficiencies, but then a liquid stream must be disposed of. Fortunately, this is no problem, for this is relatively clean water.

Noncondensing gas furnaces heat air directly, delivering to the living space via ducting. The gas is burned in diffusion burners (which entrain air for combustion as the gas is released). The products of combustion (Figure 6.10) circulate inside the tubes of a heat exchanger. Air is forced by a fan over the tubes, then flows, once heated, to the house. The efficiency of these furnaces is in the 70–80% range.

When efficiencies upwards of 90% are sought, a condensing furnace is in order. These furnaces fit the same footprint as the noncondensing ones, but are substantially different otherwise. In the noncondensing furnace, air and combustion products flow upwards, that is, in parallel flow configuration. Condensing the water vapors requires a more efficient heat transfer configuration. The hot gas must be cooled to the extent that its temperature goes below the dew-point of the mixture. Also, the water must be easily drained. In this case, a counterflow configuration, such as shown in Figure 6.11, is used. Note that the water condenses in the region where the air is coldest.

Other differences between Figures 6.10 and 6.11 are noteworthy: combustion products must flow downwards, and so a fan is required to induce the required flow against buoyant forces that tend to push the gas upwards. Also, air and gas are premixed at the fan inlet, resulting in a different type of flame. A premixed flame tends to result in more complete combustion and is also shorter than a diffusion flame.

Boilers can deliver steam or hot water (Figure 6.12). For hot water, a circulation system consisting of pumps is necessary. However, the burners and natural draft exhaust are entirely similar to those of the noncondensing gas furnace. Water circulates inside tubes and is heated in the process. The outside of the tubes is finned to ensure proper heat transfer from the gas to the water. Controls ensure that the gas

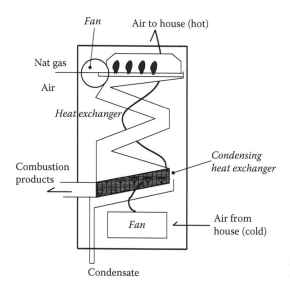

FIGURE 6.11 Gas furnace, condensing.

valve activates the gas flow only when water is circulating inside the tubes. Steam boilers have a steam drum where water is stored. In most cases, natural convection is employed to keep the water circulating in the loop. This is easy to understand, if one considers a U tube filled with liquid. The liquid column height will be the same for both legs. If one of the legs (say the right one) is heated to the point that boiling ensues, then the average density in the left leg will exceed that of the right leg, and liquid will flow to the right one. Continued liquid input to the left leg results in continual withdrawal of vapors from the right one.

For condensing boilers, similar modifications to condensing gas furnaces have evolved. Typically, a fan is used to produce premixed combustion and to force the products through finned tubes. Counterflow must be used to ensure higher efficiencies. Note that noncondensing boilers have efficiencies ranging from 78 to 88%. Condensing boilers can accommodate low return water temperatures, which raises

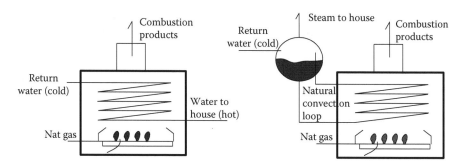

FIGURE 6.12 Hot water and steam (right) boilers.

the efficiencies to the 90 to 99% range. The cooler the return water, the cooler the gas exhaust will be, and the higher the efficiency will climb.

It is a long-standing opinion, to which this text adheres, that natural gas is too precious a fuel to use in order to deliver the relatively low temperatures of residential heating. From this perspective, gas turbines do, and fuel cells could, use natural gas more efficiently to produce power. The waste heat from those (or Rankine cycles) could then be used in district heating applications where possible.

COMMERCIAL

In magnitude, the commercial sector (office buildings, malls, education, and government) consumes a little bit less energy than the residential sector, but not by much. Figure 6.13 shows the consumption by energy types, and whereas electricity satisfies lighting, office equipment, elevators and the like, and air conditioning, the other sources are for heating. The EIA (Energy Information Agency) calculates the amount of fuel needed to produce electricity. This amount, called primary electricity, is included (Figure 6.13) as a reminder that electricity embodies a considerable component of fossil energy.

COMMERCIAL TECHNOLOGIES

Regarding basic technologies, we have little to add to the description of residential technology. The capacities and scope of the controls certainly differ, but the concepts are the same. Electric motors tend to be three-phase in large-scale applications, meaning that the stator field is generated with at least three coils, each connected to a different phase.

In a few cases, though, gas turbines have been used with natural gas to generate power and use the waste heat for heating or cooling via absorption technology. Identically, fuel cells have been installed in a few demonstration projects, generating

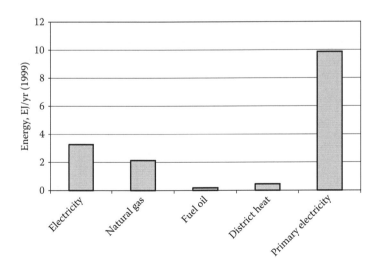

FIGURE 6.13 Commercial energy use.

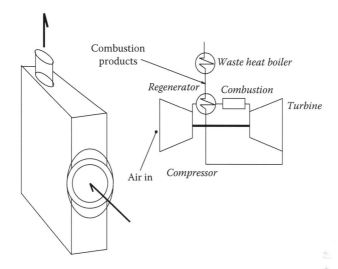

FIGURE 6.14 Microturbine, packaged and cycle components.

power and heat. In Figure 6.14, we show a schematic of a microturbine, similar in principle to the gas turbine described in Chapter 4. In these devices, heat is recovered from the exhaust gases to preheat the compressed air going into the combustion chamber. In cogeneration applications, the heat embodied in the exhaust gases is fed to a boiler or to a cooling machine.

In the case of a fuel cell, such as a phosphoric acid one, the fuel is hydrogen, and the oxidant is oxygen. Of course, to obtain hydrogen from natural gas requires a separate operation called reformation. In this operation, natural gas in a reactor with steam and oxygen can be converted into carbon dioxide and hydrogen. The former is released to the atmosphere, and the latter is used in the fuel cell. All fuel cells use the fact that hydrogen and oxygen will react completely and rapidly after a small input of activation energy. If they did not, our planet would be quite different since oxygen and hydrogen react to produce water. Water is oxidized hydrogen, and the chief reason why hydrogen is not an energy source is that it is only found in the oxidized state on this planet. In any case, hydrogen will share its electrons with oxygen, forming a covalent bond. However, the electrons will spend much more time in the proximity of the oxygen than the hydrogens of a water molecule because it is the oxygen that attracts them most. In any case, this is what happens in a fuel cell (Figure 6.15): there is an anode, there is a cathode, and there is a separator, which may be one of several but in our case is an electrolyte called phosphoric acid. As hydrogen enters the fuel cell, a catalyst (Pt) makes it lose its electron.

The electron and the ion now part ways: the electron travels towards the oxygen via a load, producing power. The ion travels through the electrolyte, and two electrons, two ions, and one oxygen react to produce water in the cathode. The electrons could shortchange the process and travel through the electrolyte, but this is impossible from a conduction viewpoint. The electrons, then, produce the current required to extract power.

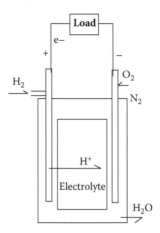

FIGURE 6.15 Fuel cell schematic.

In theory, a microturbine, working in a cyclic fashion, will produce more waste heat than a fuel cell, which is a direct energy converter. Both technology types were discussed in Chapter 2. In thermal calculations, actual efficiencies must be considered carefully, for the fuel cells have been the subject of some overestimates. Also, in a microturbine, the waste heat is available from the exhaust gas, whereas in the fuel cell a stream of steam must be cooled, condensed, and disposed of with pumps. In any case, simultaneous use of the output power and heat of these two technologies only makes sense when the user demand for power and heat coincide. In general, modern cogeneration plants using gas turbines tend to vastly exceed the electrical efficiency of either microturbines or fuel cells.

INDUSTRIAL

According to the EIA [1], industrial energy use can be classified by type of energy and end use. We plot here what falls into those categories; a substantial amount of energy, classified as other, is not shown, but it is large (6.7 EJ/yr). In any case, the energy used by industry that falls in the classifications is shown in Figure 6.16.

Natural gas plays a crucial role in industry, particularly as feedstock. Electricity (net, not primary, which is larger by a factor of three) is the next fuel, with coal a clear third. The myriad of industrial processes that consume this energy is such that a description of each would take many textbooks. We attempt here to identify some of the major users, along with a brief process description. Electric motors, gas turbines, boilers, HVAC equipment, and lighting have already been described, and their applications should be similar.

Some of the industries that consume large amounts of energy and their 2002 consumption are shown in Figure 6.17. Clearly, petroleum and coal are heavy users, and this could be labeled the energy cost of energy. Chemicals, metals, and paper follow next, with food and minerals close behind. Computers and transportation are

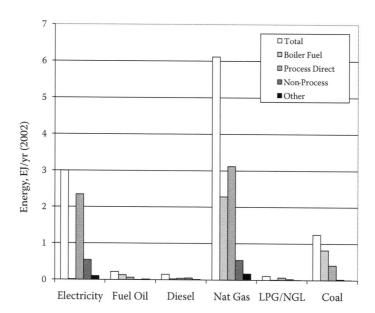

FIGURE 6.16 Industrial energy use.

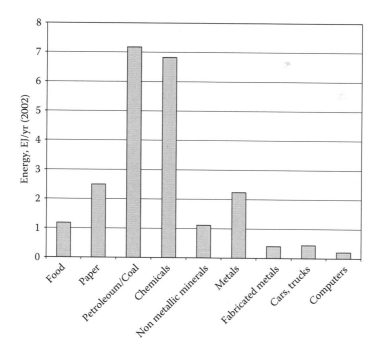

FIGURE 6.17 Energy consumption of selected industries.

shown for reference: car/truck manufacturing consumes 2.15 times as much energy as computers.

Concerning the oil/coal energy consumption by the industry, it is clear that it increases in relative terms as less oil is left. Ratios of 100 barrels out per barrel invested in the 1940s have come down to ratios of 10 to 1, as production fields and refineries cope with increased amounts of heavier oil left in the ground. Almost all of the energy consumed as given in Figure 6.17 for petroleum is in refineries, the principle of which was explained in Chapter 4. In the primary metals category, the production of iron and steel contributes 1.38 of the 2.24 EJ. Hence, we provide a summary description of the steel-making process, although the United States is not the major steel producer of the world.

STEEL

Iron mineral is mostly found as iron oxide (Fe_2O_3) in the ground. In other words, oxygen has made a covalent bond, essentially capturing all the outer electrons of the iron atoms. Of course, the oxygens are as willing to surrender their advantage as objects to fall upward or time to flow backward. They will not surrender spontaneously the electrons shared with the iron. Thus, reducing iron, make no mistake, takes energy. The basic desired reactions are essentially of reduction of the iron oxide to iron, plus other reactions to purify the product.

In order to do this, coke, iron mineral, and limestone are fed at the top of a blast furnace [6]. We follow the load (in our imagination, that is) on its way down. The most common reaction at the top of the furnace is (Figure 6.18)

$$3\, Fe_2O_3 + CO \rightarrow 2\, Fe_3O_4 + CO_2$$

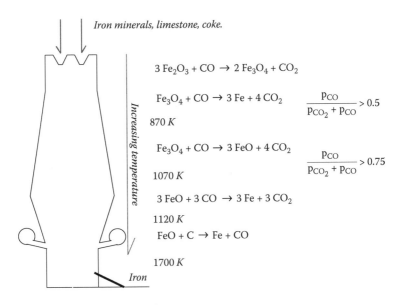

FIGURE 6.18 Schematic of a blast furnace.

Since the ratio of Fe to O has increased, one can say that reduction has already started at the top of the furnace. However, from magnetite there is still a long way to Fe. The next reaction sought takes place below 840 K:

$$Fe_3O_4 + CO \rightarrow 3\,Fe + 4\,CO_2$$

as long as the ratio of partial pressures is such that

$$\frac{p_{co}}{p_{co_2} + p_{co}} > 0.5$$

Also, in order for the reaction to proceed, a continuous supply of CO must be ensured. Above 870 K, two successive reactions must take place for reduction to continue:

$$Fe_3O_4 + CO \rightarrow 3\,FeO + 4\,CO_2$$

$$3\,FeO + 3\,CO \rightarrow 3\,Fe + 3\,CO_2$$

An abundant supply of CO is necessary for these reactions to occur in the desired direction. As the temperature increases, so must the partial pressure of CO. In fact, above 1070 K, the constraint is

$$\frac{p_{co}}{p_{co_2} + p_{co}} > 0.75$$

Another reduction that takes place near the bottom of the furnace is between the coke and the ferrous oxide still left over:

$$FeO + C \rightarrow Fe + CO$$

For this reaction level, the temperature must exceed 1120 K. The molten iron is collected at the bottom of the furnace and exhausted. Air is injected at the bottom of the furnace, and gases flow upwards and exhaust at the top. We follow the gases, again, in our imagination. The energy for all these reductions comes from the combustion of coke. At the air injection point, we have

$$O_2 + C \rightarrow CO_2$$

The CO_2, when excess carbon is present (found on the way up), produces

$$CO_2 + C \rightarrow 2\,CO$$

which is exactly what is needed. The decomposition of the lime also generates carbon dioxide, and it is useful for the removal of impurities (such as sulfur) along with the slag.

Once iron is obtained this way, steel production requires further control of the chemical composition and reduction of the carbon content. The iron is called pig iron, and it must be modified to obtain steel. The crucial modification is to reduce

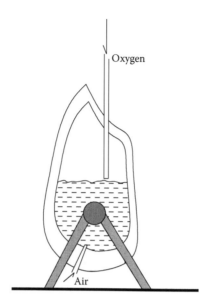

FIGURE 6.19 Generic converter.

the carbon content, an operation typically done by injecting oxygen to remove the carbon as carbon monoxide and dioxide. The iron is oxidized by the oxygen, and then it is reduced by the carbon via typical reactions such as

$$C + OFe \rightarrow CO + Fe$$

$$CFe_3 + OFe \rightarrow CO + 4\,Fe$$

Note that not all of the iron is oxidized, just what is in the proximity of the oxygen. However, as shown by the preceding reactions, carbon leaves the melt as carbon monoxide (and also dioxide). Several configurations exist to carry out this refining of iron into steel, with suitable alloying elements being introduced and purification reactions taking place. The Bessemer or Thomas converters, and some others, tend to share a shape resembling the one shown in Figure 6.19. Air is introduced from below to refine the iron, and this is quite a sight, for a large flame projects from the converter's mouth during the blowing operation. Oxygen injection with a lance is less impressive, but it is effective in keeping the metal molten during oxidation, using precisely the heat released by the reactions. Similar practices take place in other converters that essentially have a pool that is drained after conversion.

Electric furnaces produce high-quality steel from metal scrap, and their use is on the rise. In these furnaces, oxygen is used only occasionally. How does the carbon oxidation take place? Simple: the charge can include some iron mineral. A current is circulated through the charge to melt it and keep it molten. Electrodes (Figure 6.20) made of graphite or carbon agglomerates sustain a voltage difference sufficient to

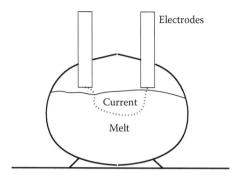

FIGURE 6.20 Electric arc furnace.

maintain a current through the melt, which arches from the melt to the electrodes. The sought-after reaction is

$$2\,C + O_2 Fe_3 \rightarrow 2\,CO + 3\,Fe$$

and the reduction of the mineral implies the production of additional steel. Of course, the high and uniform temperatures, plus the lack of oxygen, favor purification reactions everywhere with suitable compounds added to the charge leading to low sulfur and phosphorus content. Also, alloying elements do not oxidize and stay in the melt. A slag floating on the top tends to concentrate impurities. In short, a more uniform steel of controlled composition is obtained with this technology. Of course, the electrical current requires energy. About 1.5 kJe/gr (440 kW · hr/ton) [7] seems to be needed for steel production. The melt is discharged from the furnace and solidified in ingot molds.

Aluminum

Continuing with primary metals, reference must be made to the production of aluminum (Al). Although Al is found in abundance all over the world as bauxite, {Al$_2$O$_3$, n H$_2$O} (aluminum oxide complexed with water, with n = 0 to 3), the oxygen must be convinced to return the electrons to the Al atoms, a difficult task given the electronic affinity of O. Clay is another, noncommercial, source of Al, abundant in the United States, but it has little bauxite. To produce the metal from imported bauxite, much energy is needed.

The Bayer process converts the bauxite into alumina, Al$_2$O$_3$, that is, it gets rid of the water. How this happens is described with variations, and we offer here a synopsis from [8,2]. The bauxite is attacked with sodium hydroxide to yield a soluble aluminate of sodium. The key here is solubility, for other metals (iron, titanium) produce insoluble compounds. The soluble aluminate is hydrated, giving aluminum hydroxide. We have then two reactions:

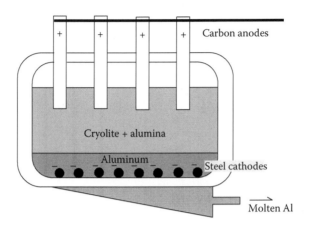

FIGURE 6.21 Electrolysis of alumina to yield aluminum.

Solubilization

$$Al_2O_3 \, nH_2O + 2\,NaOH \rightarrow 2\,AlO_2Na + (n+1)\,H_2O$$

Hydration

$$AlO_2Na + 2\,H_2O \rightarrow Al(OH)_3 + NaOH$$

The NaOH is separated by decanting and reused. The aluminum hydroxide is converted into alumina by heating to 1200°C, as follows (calcination):

$$2\,Al(OH)_3 \rightarrow Al_2O_3(s) + 3\,H_2O$$

With a pure aluminum oxide in hand, the process that consumes the most energy can begin. The electrolysis process is described with the help of Figure 6.21. In the cathode, the aluminum is reduced:

$$Al^{3+}(melt) + 3e \rightarrow Al(l)$$

The carbon anode participates in the reaction in such a way that it is consumed:

$$2\,O^{-2}(melt) + C(s) \rightarrow CO_2(g) + 4e$$

That is, the carbon anode goes to carbon dioxide. For these reactions to occur, the aluminum and oxygen ions must have mobility, namely, the aluminum oxide (Al_2O_3) must be in the liquid state in such a way that it ionizes. The problem is, the melting temperature of Al_2O_3 is 2323 K, and a rather complex problem presents itself: what to use to hold the molten metal, and how to insulate it thermally. Another problem is that the melt is not conductive, that is, some other substance must be added to result in separation. To overcome these difficulties, cryolite is used. About 13% alumina is added to a bath of cryolite (Na_3AlF_6) that melts at 1223 K, and the electrolysis can proceed, for the bath is now conductive.

PAPER

Paper mills consume 1.06 EJ/yr as of 2002. The manufacture of paper is energy intensive, although much has been done to use renewable energies from wood residues in various forms. A brief description of the paper mill technology is offered here.

For our purposes, wood, the main raw material, can be thought of as fibers made up of cellulose, embedded in a mass of lignin and hemicellulose and other compounds. Cellulose is a polymer of glucose, and the fibers are what largely make up the paper, although we still have a long way to go from fibers to paper. Starting with wood, it must be harvested and chipped. Then, the fibers and lignin must be separated from each other. The process is summarized in Figure 6.22.

Two main processes for separation have been developed: digester and thermomechanical (TMP). The objective of the digester is to dissolve the lignin. A solution (white liquor) of sodium hydroxide (NaOH), sodium sulphide (Na_2S), and the wood chips is cooked for a few hours with continuous steam addition under pressure. When the mixture pressure is released, the dissolved lignin (black liquor) and the fibers are separate entities. The fibers are separated by washing and filtration, and stored or sent to paper making. The black liquor is reconcentrated by distillation and burned for steam production. The boiler collects spent chemicals, which are regenerated into white liquor to restart the cycle.

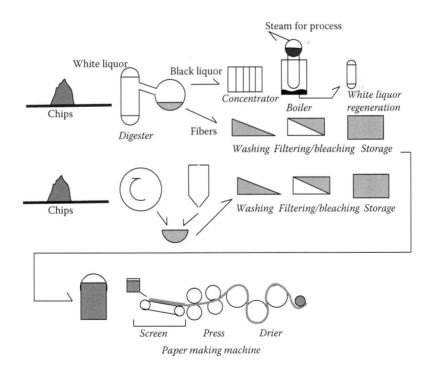

FIGURE 6.22 The paper-making process.

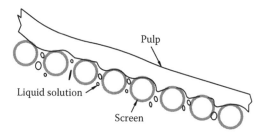

FIGURE 6.23 The pulp is concentrated in the fast-moving screen.

The TMP (Figure 6.22) process produces fiber that is not as high quality as that produced by the digester (or Kraft) process. Wood chips are fed between two spinning disks. The friction thus created converts the mechanical energy fed to the disks into heat (talk about entropy!). The wood chips explode into lignin and steam and fibers. The fibers are separated via gravity and cyclones, and collected for paper making after washing and filtering.

The fibers are bleached further with chlorine dioxide (ClO_2) or hydrogen peroxide (H_2O_2) or O_2, and suspended in an aqueous solution in the box of the paper machine. What follows is quite a sight. The solution flows into a fast-moving screen (Figure 6.23).

Some of the liquid falls through the moving screen, but the fibers form a bed of uniform thickness in the traverse direction but decreasing in the direction of screen motion. This happens fast: speeds are on the order of 100 to 120 m/s. The fibers are now pressed between a number of rollers to decrease the moisture content further to 40–50%. The rest of the water is removed by making the continuous sheet move over rotating cylinders heated internally with steam. Each cylinder rotates at a different speed because, as the paper dries, it shrinks, and in the absence of different speeds, the paper would rupture.

Instead of wood chips, recycled paper can be used as feedstock. In this case, the paper is reduced to pulp without digesters or TMP processes simply by suspending the fibers in water after dissolving the ink. Typically, new fiber and recycled fiber are blended to obtain the desired quality of paper. As per the EIA, the energy consumed to produce one 1992 dollar worth of paper in 1994 was 20.8 GJ [9].

PLASTICS

Plastics manufacture requires significant energy: 1.92 EJ per year. Plastics are polymers, the building blocks of which come from hydrocarbons. Thermal processes of oil (decomposition) and natural gas result in ethylene (ethene, C_2H_4), propylene (propene, C_3H_6), and butene (C_4H_8), among others. These are called monomers. Monomers (Figure 6.24) are used as building blocks for polymers.

There are two types of plastics: thermosets and thermoplastics. The former will not soften with increased temperatures, whereas the latter will. Additives are used

H   H   H−C−H
  \\ / |
   C   H−C−H
   ‖ |
   C   H−C−H
  / \\ |
H   H   H−C−H
                |

Monomer   Polymer

FIGURE 6.24 Ethylene (ethene) monomer and corresponding polymer (polyethylene).

to give them different properties. Since the feedstock is essentially oil or natural gas, it is clear that plastics have a large energetic component. The polymerization is exothermic, but injection machines melt the plastic with consumption of electrical energy in most cases.

A flowchart shows several steps to obtaining polyethylene (Figure 6.25). The distillations, compression, and cracking have a net energy input, whereas the other operations release energy or have no substantial energy input. The light hydrocarbons (3 to 4 carbon atoms) are mixed with steam and rapidly heated to between 1070 and 1120 K, flowing over the hot tubes of a furnace. Reactions take place that produce some double bonds and shorter hydrocarbons. The gas is cooled rapidly, and a compression/distillation process follows. The distillation yields the desired ethene. In a reactor with catalysis to avoid large temperatures (that is, to allow the reaction to proceed at low temperatures), the ethene combines with other ethenes to form

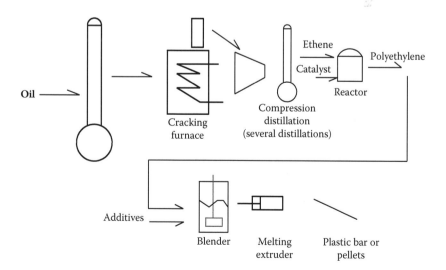

FIGURE 6.25 Flowchart for plastic production.

polyethylene, which leaves as a powder. Blended with additives, the polyethylene is now melted and extruded to form bars that can be cut into pellets.

TRANSPORTATION

Cars, some brands more than others, are undoubtedly an engineering marvel. We briefly described the IC engine in Chapter 4, but there are so many other subsystems that it would take a long time to learn about them all, and much more time to design them. Trucks, buses, airplanes, and trains also have complex subsystems. The amount of practical design experience and years of manufacturing embodied in each car is large. Hence, we do not describe the technology here but how much energy the transportation sector uses, as given in Reference 10 for 2002. More recent data are available on the same site, and the reader is encouraged to validate the conclusions that follow.

First, we focus on total energy consumption (Figure 6.26). Clearly, cars and pickup trucks consume most of the transportation fuels, with aircraft not far behind. The other modes barely make a difference, even considering that our source does not make it clear whether energy used by rail commuter is primary or direct electricity. In any case, reductions in the transportation energy budget could come from more efficient IC and gas turbine engines.

Overall efficiency numbers show that increased mass transit ridership could do wonders for energy consumption, not to speak of the social life of many of its riders. In any case, consider energy per vehicle per kilometer (Figure 6.27).

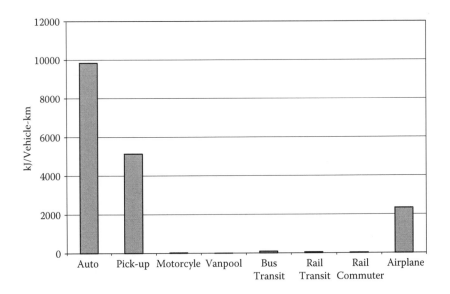

FIGURE 6.26 Total energy consumption in transportation (selected modes, 2002).

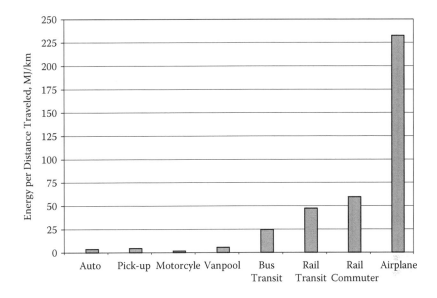

FIGURE 6.27 Energy per vehicle/distance, 2002.

Clearly, it takes a lot of energy to move an airplane, which has to move fast to stay aloft, and much less energy to move a car, with all the other modes in between. However, consider that each transportation mode can carry a different number of passengers, and that the number of passengers is much greater in a train than in a pickup truck.

Figure 6.28 shows the energy per passenger-km. The conclusions are rather staggering: vanpooling is the more efficient transportation means when it boils down

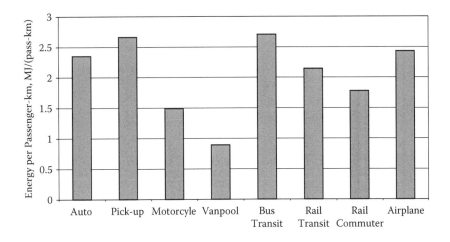

FIGURE 6.28 Energy per passenger-kilometer.

to energy per passenger-km. All other modes are somewhat similar, reflecting the utilization factors assumed by the source of data [10]. Only if ridership in bus and rail transit increased (i.e., the utilization factor of those means of transportation increased) would they become competitive with vanpools or other means of mass transportation. As fuel prices increase, such a process is widely reported by the media to be taking place.

When energy per passenger numbers are considered, the airplane becomes competitive with virtually every other mode except vanpools.

EXAMPLES

We present transient simulation of a small building. The idea is to study thermostat setback as a means of energy savings.

Example 6.1 (VisSim©), Building Parameters, develops a simple model of a building to use in transient calculations.

EXAMPLE 6.1 A SIMPLE MODEL OF A BUILDING FOR HEAT LOSS CALCULATIONS [VISSIM]

To approximate the performance of a building in winter, it is necessary to model the heat flow from the inside to the outside of the building (Figure Example 6.1.1).

Even though the aesthetics of our building could not support any awards, it is clear that, as a thermodynamic system, all buildings (ours included) must contend (in winter) with the fact that thermal energy flows from hot to colder temperatures. The heating system must supply energy at the same rate as it is lost, or approximately so, to keep the temperature constant. We show in this example how one can approximate the heat loss calculations and the time constant of a building.

The inputs in the example are the thermal resistance values of walls, ceiling, and either floor (aboveground construction) or basement. Thermal resistance measures the resistance to heat flow across the element (wall, ceiling, etc.) in such a way that

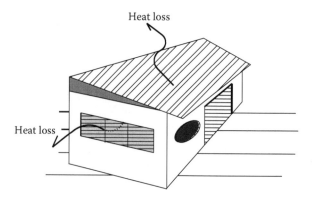

FIGURE EXAMPLE 6.1.1 A simple building.

the heat flow as given by Equation 2.25 (Chapter 2):

$$\frac{dQ}{dt} = k \cdot A \cdot \frac{(T_1 - T_2)}{\Delta x} \tag{2.25}$$

is cast as

$$\frac{dQ}{dt} = A \cdot \frac{(T_1 - T_2)}{\dfrac{\Delta x}{k}} \tag{a}$$

The term $(\Delta x/k)$ is the resistance of the material that makes up the wall, namely,

$$R_{th_i} = \frac{\Delta x}{k} \tag{b}$$

whereby the resistance is proportional to the material thickness and inversely proportional to the thermal conductivity. In home improvement and hardware stores, insulation is rated by its R-value in the units $(ft^2 \cdot F/BTU/hr)$. Naturally, more than one material makes up the wall, and hence, the values specified in the example reflect the combination of materials and convective coefficients, rather than just simply insulation. Trade and government organizations publish desired (or mandated) minimum R-values for residences.

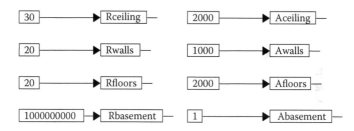

The huge resistance values for the basement reflects its absence. The areas for each surface are included with this input. Windows are rated differently from walls for winter performance. The performance parameter is the U-factor, in [BTU/hr/ $(ft^2 \cdot F)$].

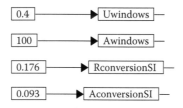

The two conversion factors are used to convert R and U values to SI units. Also, the U-factor is the inverse of the R-value, and hence in the block

$$\boxed{\text{Unit Conversion}}$$

the R-values (including 1/U for windows) are converted to the units $[K \cdot m^2/W]$. The total heat loss is obtained by adding the heat loss for each surface and assuming the same indoor and outdoor temperature. Thus, from Equations (a), (b) extended to each surface i and adding up all the contributions, one obtains

$$\frac{dQ_{total}}{dt} = \sum_i A_i \cdot \frac{(T_1 - T_2)}{R_i} = (T_1 - T_2)\sum_i \frac{1}{(R_{th_i}/A_i)} = \frac{(T_1 - T_2)}{Ro} \qquad (c)$$

where Ro is the overall resistance of the building, and it is calculated as

$$Ro = \left(\frac{1}{\sum_i \dfrac{1}{(R_{th_i}/A_i)}} \right) \qquad (d)$$

with units of $[K/W]$. This calculation is effected in block

$$\boxed{\text{Overall Resistance, Ro K/W}}$$

The house heat loss, under steady-state conditions (i.e., constant indoor and outdoor temperatures) is given by Equation (c) in block

$$\boxed{\text{Heat Loss, W}}$$

with the result that this building loses heat at a rate as shown in

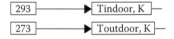

That is, every second at the given temperatures, the building loses at the rate of 2.7 kW, which is really not much. Over 10 hours, this load entails $2.7 \cdot 10 = 27 \text{ kW} \cdot \text{hr}$, or a cost of \$2.70. Additional heat losses due to infiltration and ventilation are not considered here.

Example 6.2 (VisSim), Building Thermal Capacity, shows the time constant of a small building.

EXAMPLE 6.2 THE BUILDING TIME CONSTANT [VISSIM]

In Example 6.1, the overall resistance Ro to heat transfer of the building was introduced. Another useful concept is the thermal capacity C of the building. In conjunction with Ro, it establishes how fast the temperature of the building will decrease with time. Modeling the transient behavior of a building is good practice to develop energy conservation strategies and comprehend what heating plant controls must accomplish. Hence, we develop a generic (simple) model of a residence or small building aimed at calculating how the building temperature (Figure Example 6.2.1) will evolve with time.

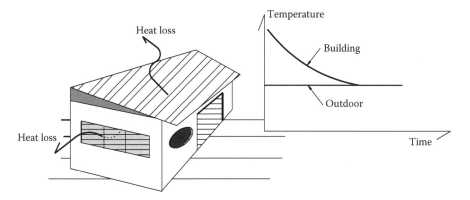

FIGURE EXAMPLE 6.2.1 Temperature decreases in the absence of heat input.

For this purpose, we apply the First Law to the control volume (CV) surrounding the building at any time. We only focus on energy here, not mass.

$$\frac{dQ}{dt} - \frac{dW_\sigma}{dt} = \frac{dE_\sigma}{dt} + \dot{m}_o \cdot \left(h + \frac{V^2}{2} + g \cdot z\right)_o - \dot{m}_i \cdot \left(h + \frac{V^2}{2} + g \cdot z\right)_i \qquad (2.14)$$

Whereas mass indeed leaves and enters a building in the form of people, air, and water among other things (coffee, groceries, etc.), we ignore them here, although we note that infiltration losses are generally well characterized. In principle, everything could be accounted for, but numerical modeling requires criteria regarding which inputs/outputs to include. Hence, focusing solely on energy, we have

$$\frac{dQ}{dt} - \frac{dW_\sigma}{dt} = \frac{dE_\sigma}{dt}$$

We assume that other than the heat loss due to conduction calculated in Example 6.1, no other heat input or output exists. Then, the foregoing equation reduces, with T_o the outdoor temperature, to

$$\frac{dQ}{dt} = \frac{dE_\sigma}{dt} = C_{blg} \cdot \frac{dT}{dt} = -\left(\frac{T - T_o}{Ro}\right) \qquad (a)$$

where C_{blg} is the thermal capacity of the building and T its temperature. The value of C_{blg} includes the thermal capacity of everything inside the building. Living things have mechanisms to keep their temperatures nearly constant, and hence, they contribute to C_{blg} in a different fashion than, say, a table would. However, we neglect those contributions for the sake of simplicity, not because they might be negligible. For instance, in a tent (minimal C_{blg}), they would not be negligible. In any case, we have

$$\frac{dT}{dt} = -\left(\frac{T - T_o}{Ro \cdot C_{blg}}\right) \qquad (b)$$

that is, the building temperature will decrease as long as $T > T_o$. Although Equation (b) can be solved analytically, we choose instead to solve it numerically in the VisSim Example 6.2. In this example, Equation (b) is solved in the block

$$\boxed{\text{Eq (b)}}$$

with the data

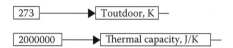

and the same Ro value of Example 6.1. The blocks to solve Equation (b) are

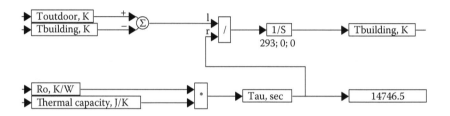

and it should be noted that the product $Ro \cdot C$ has the units of time and is called Tau. It represents a time constant close to four hours. It can be shown that 67% of the original temperature difference disappears in the time Tau. In any case, the temperature versus time (for an initial building temperature of 293 K) looks as shown in Figure Example 6.2.2.

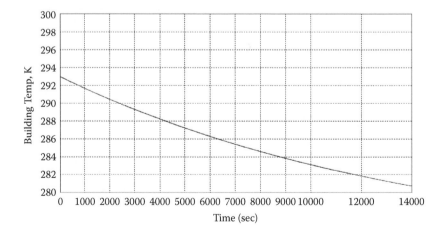

FIGURE EXAMPLE 6.2.2 Calculated temperature decrease in the absence of heat input.

It is clear that the rate at which the temperature decreases becomes smaller as time flows. If it is desired to reverse this trend, a heat input must be provided. How to control the rate when heat input is added is the topic of Example 6.3.

Example 6.3a and b (VisSim) Building Temperature Control, shows how ON–OFF and proportional–integral controls can be used in buildings.

EXAMPLE 6.3 A,B CONTROLLING THE BUILDING TEMPERATURE AND SETBACK [VISSIM]

In Example 6.2, there was no heat input to the building. The building now has a stack, indicating that a heating plant furnishes a heat input to the building. The building temperature will depend on the capacity of the heating plant, the outdoor temperature, and the type of control adopted to control the heat input.

For instance, as shown in Figure Example 6.3.1, the building temperature (saw tooth) will vary with time by increasing when the heat input is activated, and decreasing when it is absent. The setpoint indicates the desired temperature as set in a thermostat. The heat is activated when the temperature drops below the setpoint and reaches the setpoint minus half of the deadband. After the heat is activated, the temperature rises until the setpoint plus half deadband is reached and the heat is turned off. During a period of no occupancy or perhaps at night, the setpoint is decreased by Δ, the setback. Other control strategies are possible, and they are outlined as follows.

To model this process, we start with the energy balance for the building, that is, we apply the First Law to the CV surrounding the building. We only focus on energy here, not mass.

$$\frac{dQ}{dt} - \frac{dW_\sigma}{dt} = \frac{dE_\sigma}{dt} + \dot{m}_o \cdot \left(h + \frac{V^2}{2} + g \cdot z \right)_o - \dot{m}_i \cdot \left(h + \frac{V^2}{2} + g \cdot z \right)_i \qquad (2.14)$$

Whereas mass indeed leaves and enters a building as people, air, and water among other things (coffee, groceries, etc.), we ignore them here, although we note that air infiltration losses are generally well categorized in load calculation procedures. In principle, everything could be accounted for, but simplicity demands

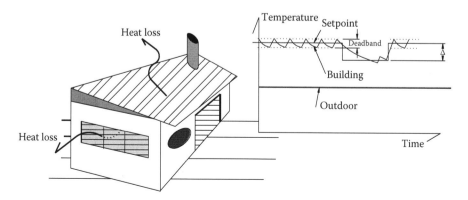

FIGURE EXAMPLE 6.3.1 A possible way to control temperature.

criteria regarding which inputs/outputs to include. Hence, focusing solely on energy just as in Example 6.2, we have

$$\frac{dQ}{dt} - \frac{dW_\sigma}{dt} = \frac{dE_\sigma}{dt}$$

In winter, heat must be supplied, typically by combustion of a fuel. Hence, the above equation, with T_o as the outdoor temperature and R and C_{blg} the resistance and thermal capacity of the building, reduces to

$$\frac{dQ}{dt} = \frac{dE_\sigma}{dt} = C_{blg} \cdot \frac{dT}{dt} = -\left(\frac{T - T_o}{Ro}\right) + Q_{in}$$

which allows us to arrive at

$$\frac{dT}{dt} = -\left(\frac{T - T_o}{Ro \cdot C_{blg}}\right) + \frac{Q_{in}}{C_{blg}} \tag{a}$$

that is, the building temperature is a balance between the heat loss and the heat input produced by a furnace or heat pump. Of course, electric inputs (such as from incandescent lighting) can be considered and are part of any heat load calculation, but we ignore them here for our example. Note that both T_o and Q_{in} are functions of time, so that Equation (a) is only "seemingly" simple.

The outdoor temperature is assumed to vary as a sinusoid with a period of one day (86400 s), around a mean of 273 K (0°C, 32°F) (Figure Example 6.3.2).

The mean outdoor temperature is an input that can be changed by the user. The sinusoidal variation could be modified as well, having been adopted just to show how a variable outdoor temperature can be handled. The amplitude is 10 K, quite high,

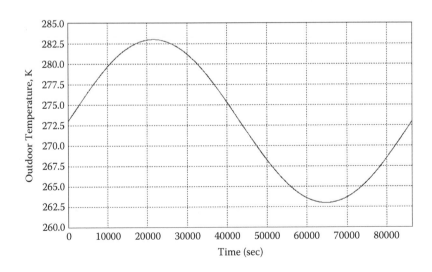

FIGURE EXAMPLE 6.3.2 Outdoor temperature variation.

and the reader might argue that such amplitudes may only occur in desertic areas removed from large bodies of water. The block "outdoor temperature" models the assumed temperature profile (Figure 6.3.2).

The heat input to the building, under an ON–OFF control, is not continuous but active only when the building temperature is rising, and (a) either below the deadband lower limit, or (b) within the deadband. In those instances, the heating plant delivers its full capacity. Equation (a), then, can be integrated with a heat input equal to the heating plant capacity:

$$\boxed{\text{Eq (a) ex 6.3}}$$

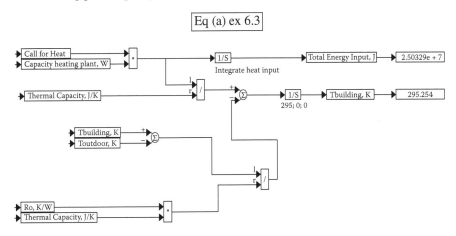

In the foregoing blocks, the differential equation for "T building, K" (Equation (a)) is solved for a heat input equal to the "Capacity of the heating plant, W." This capacity (specified by the user) is only active when the variable "Call for heat," which can take the values 0 or 1, is equal to one. The control action to activate the heating input (i.e., to make "Call for heat" equal to one), is implemented in the block

$$\boxed{\text{ON-OFF Control}}$$

In this block, use is made of a relay block in VisSim, as follows.

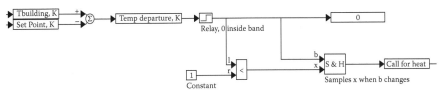

The temperature difference between the setpoint and the building temperature is the "Temp departure K." This variable inputs a relay with output zero when the signal is within the deadband, or −1 when below its lower limit, or 1 elsewhere. When the relay output is smaller than one, the output of the Boolean block is one. This output is the x input to an S&H (Sample and Hold) block. When b changes (i.e., when the temperature departure crosses either deadband limit), the Boolean output is sampled. The output of S&H equals input x and becomes the "Call for heat" variable. We test a few simple cases to better explain the logic.

TABLE EXAMPLE 6.3.1
ON–OFF Control Values

T building (K)	Setpoint (K)	Temp departure (K)	Call for heat
290	290	0	0
285	290	−5	1
295	290	5	0

The test values are shown in Table Example 6.3.1. Additionally, the reader may want to notice that, as the relay switches from −1 to 0, the Boolean block output remains equal to 1, that is, heat will be provided as long as the temperature is below the upper deadband limit.

The setpoint can be used to save energy by reducing the building temperature at night. The setback (Δ in Figure Example 6.3.1) is activated between time periods chosen by the operator.

Setpoint change timing

Once "setback begin" and "setback end" points are specified, the output of the summing junction is −1 in between the two points in time, and a negative Tsetpoint change is generated. The change is used to adjust the setpoint in the block "ON–OFF control." The results of our model are easily interpreted via Figure Example 6.3.3.

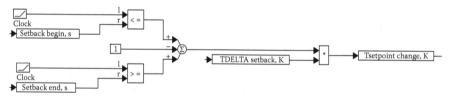

The building temperature is maintained within the deadband in a sawtooth pattern, ascending when the heat is on, descending when it is off. As the outdoor temperature decreases at night, the setback is activated, and the building loses heat at much lower rates than without setback. The total energy input is with a setback of 12°K, suitable for an unoccupied building.

For a smaller setback of say 6°K, the energy consumption in the same building would be as shown in

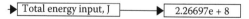

Whereas, with no setback, the energy input would be as shown in

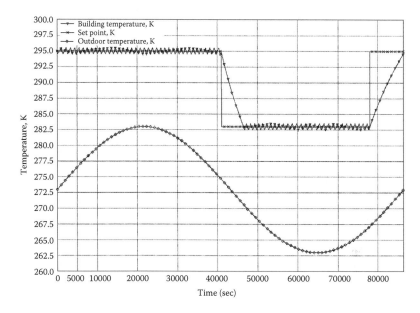

FIGURE EXAMPLE 6.3.3 ON–OFF control with setback.

If instead of ON–OFF controls a proportional–integral control is employed, the temperature traces of Figure Example 6.3.4 arise. In Example 6.3b, a PI controller is implemented in lieu of the ON–OFF control. The building temperature trace (Figure Example 6.3.4) shows that the control results in values close to the setpoint. However,

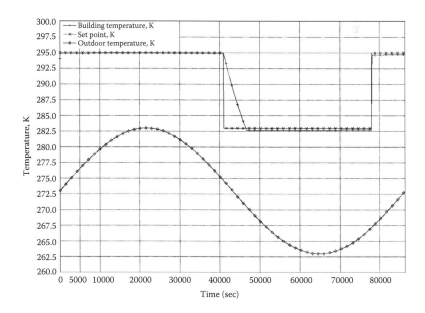

FIGURE EXAMPLE 6.3.4 PI control of building temperature.

modulating equipment is required for this application. This type of equipment is slowly appearing in the residential market.

The details of this PI model can be obtained from Example 6.3b in VisSim on the CD.

SUMMARY

Perhaps the best summary was the quote of Louis Aragon below this chapter's heading. End use applications are indeed only limited by the imagination of countless innovators throughout the world. Nevertheless, we attempt here a less ambitious synopsis. The residential sector uses considerable amounts of fuel and electric power, not always efficiently. Typically, higher efficiencies entail higher initial cost, and a host of reasons preclude the acquisition of high-end equipment by all. Perhaps in a not-too-distant future, energy conserving equipment will indeed be introduced in all or most households. Commercial end-use technology is similar to that of residential, although of sturdier construction. However, microturbines and fuel cells have mainly powered commercial demonstration projects to highlight energy sufficiency or technical prowess. The industrial sector uses energy to transform materials. The typical reaction is reduction, for metals of extensive use need to be deoxidized, an operation that requires large amounts of energy. For plastics, the energetic bonds of hydrocarbons are preserved, but their conversion into large chains also calls for energy, as does the distillation and purification of oil products. The reader will perhaps see in all these transformations the overwhelming conclusion embodied in the Second Law: nonspontaneous transformations require energy. In some cases, when the process is spontaneous, such as combustion in a boiler, it can only be sustained by fuel consumption, which is a direct, if trite, conclusion from the First Law. A remarkable exception is the production of paper via chemical processes: some of the energy embodied in the wood is recovered to yield energy for drying paper. Still, the process is not a net energy producer, as the Second Law would allow us to anticipate.

REFERENCES

1. EIA, http://www.eia.doe.gov/emeu/recs/recs2001/detailcetbls.html
2. Kinsler, M., 2000, The Y1936 problem, *Invention and Technology*, Winter 2000.
3. http://hyperphysics.phy-astr.gsu.edu/hbase/vision/lumpow.html
4. Žukauskas, A., Gavriušinas, V., Functional combinations in solid states, http://www.mtmi.vu.lt/pfk/funkc_dariniai/program.htm)
5. Otherpower.com, http://www.otherpower.com/otherpowerfront.shtml
6. Hilly, G., Chaussin, C., 1967, *Curso Basico de Metalurgia y Siderurgia*, Ed Montesó, Barcelona.
7. http://en.wikipedia.org/wiki/Electric_arc_furnace
8. Atkins, P., Jones, L., 2002, *Chemical Principles: The Quest for Insight*, W.H. Freeman, New York.
9. EIA, http://www.eia.doe.gov/emeu/mecs/mecs94/ei/ei_1.html#table 12.
10. ORNL, Oak Ridge National Laboratory, http://www-cta.ornl.gov/data/chapter2.shtml

7 Chain Efficiencies
From Capture to Utilization

To many of the futures (not to all) I leave my garden of the bifurcating paths.

Ts'ui Pen in the *Garden of the Bifurcating Paths,* **J. L. Borges**

When connecting source and end use, many processes (or only a few) can be invoked. Each series of processes configures a path. Not all paths have the same efficiency, and hence, some are preferable to others. This chapter summarizes what we could find regarding process efficiencies so as to enable the reader to formulate multiple paths of interest and to assess their efficiencies.

ON THIS CHAPTER

When facing alternative paths in a labyrinth, in an unknown garden, or in the mountains without valid clues, one's decisions as to which path to follow cannot be based on reason. One path may be just as attractive as the other, but much more dangerous, or may lead nowhere. Yet, decisions must be made. In a labyrinth or such, intuition or chance may play a role. In engineering, "we cannot leave the haphazard to chance."* In other words, we should rely on chance as little as possible, even if chance, for lack of a better concept to denote our limitations to ascertain the future, is always present. The engineering endeavor is based on reason, but the future is open in all directions, namely, it offers multiple paths.

A reasonable clue in choosing paths would have to deal with energy availability, that is, developing a technology that impacts a scarce energy supply may not be a desirable path. The technology will impact the energy supply in direct proportion to its market penetration and in inverse proportion to its average efficiency. The market penetration implies an acceptable initial price and a reasonable operating cost. The latter is again inversely proportional to the efficiency. Hence, efficiency appears as a first-order consideration in choosing a path. Efficiency also matters when it comes down to waste or undesirable byproducts, because the higher the efficiency, the less of them will be produced.

In this chapter, we endeavor to provide an approximate methodology to allow the reader to estimate the efficiency of multiple paths for energy harvesting, generation, and use. A word of warning: biomass has the unique property of producing oxygen while growing. Beyond this indispensable utility, biomass enhances the environment in ways that an electrolysis or water dissociation plant cannot. Credit is not assigned to biomass in this regard because oxygen is so vital that it needs no justification.

---

* N. Simpson in *Bent Perceptions* by W. Starr, Kinston Ellis Press.

STEADY-STATE EFFICIENCIES

The response time of technologies is fundamental to their acceptance, and it seems to be getting shorter all of the time. Steady-state efficiencies may measure some sort of average performance, but because they lack dynamics content, they only allow broad first-order conclusions regarding energy use. In any case, we present briefly the method with what little reliable data we can find. The aim is to develop an understanding of how the Laws of Thermodynamics rule energy transformations as different technologies with the same end use are adopted.

The method is conceptually simple. From fuel to some final product, a number of conversions, transport, and distribution operations must take place. Each costs energy. The idea is to calculate an overall ratio of energy output to energy input, to at least provide some guidance as to the feasibility of multiple transformations. Electricity production is a case that has already been discussed. As per Chapter 3, one unit of electric energy is equivalent to $1/0.375 = 2.67$ units of thermal energy.

Hence, a unit of electrical energy is not entirely equivalent to a unit of thermal energy, for it takes between 1.7 and 3 units of thermal energy to obtain one unit of electrical energy. Hold it, you say: if I use a 1-kW electrical heater, I can furnish 1 kW·hr in one hour, and that is pure thermal energy. Meanwhile, back at the power plant, about 1.6 to 3 kW · hr were used to heat the room. Had the room been heated directly (possible with some fuels and impossible with others), only 1 to 1.2 kW · hr would have been used. Natural gas is an excellent fuel. To produce electricity for use in space heating with resistors is a good way to waste it; but to buy into this statement, one must know the efficiency of production. Therein lies the difficulty. Folks argue intensely about these efficiencies, and agreement does not exist.

However, by taking maximum and minimum efficiency values we might reach some useful conclusions. We are guided by the best of intentions, and we certainly hope to avoid any paths, roads, highways, or superhighways leading to the uncomfortable (if not afflicted by energy scarcity) destination of proverbial roads paved with good intentions. The principle is simple, and to avoid written words that might obstruct reading, we used pictures to illustrate each conversion path. Data from the literature are used abundantly, but on occasion we bring our own experience and judgment into the fray. There are no universal principles for this evaluation, except that if something looks too good to be true, it probably is.

The method is uncomplicated. If one has process I, then there is a ratio of energy out to energy in (typically not of the same type but measured in the same units):

$$\eta I = \frac{EI_{out}}{EI_{in}}$$

A second process may take the output of I as input and use that input to produce a useful output II:

$$\eta II = \frac{EII_{out}}{EII_{in}} = \frac{EII_{out}}{EI_{out}}$$

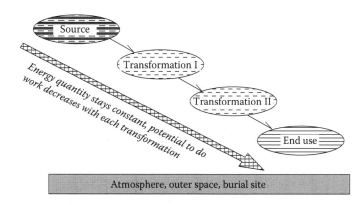

FIGURE 7.1 Overall efficiency of energy conversion processes in sequence.

so that the overall process efficiency is

$$\eta_O = \frac{EII_{out}}{EI_{in}} = \eta I \cdot \eta II$$

or, in general, for many processes, we have

$$\eta_O = \prod_i \eta_i \tag{7.1}$$

The overall efficiency captured by Equation 7.1 is illustrated in Figure 7.1. As one process inputs energy and transforms it into a different type of energy, an efficiency is defined. The First Law indicates that perfect accounting is possible, and hence, all the energy out should equal the energy input. The Second Law indicates that some energy will be lost as waste heat and that the temperature of that heat cannot be upgraded to useful values without further energy expenditure. Hence, the useful output will never equal the useful input because every transformation entails losses.

One last observation deals with the obvious fact that the efficiencies thus defined do not take into account the quality but just the quantity of energy in play. Doing so would further complicate and obscure a topic that, although uncertain, is not completely inaccessible. If the starting and ending points of comparative paths are the same, then the demands on the fuel resource will clearly reflect the projected overall efficiency. Hence, for comparative purposes with the same beginning and end, quality considerations can be obviated.

EXTRACTION ENERGY COSTS

Table 7.1 illustrates the efficiencies of extraction for fossil fuels. This is the ratio of energy made available for use divided by the sum of the available energy and the energy input to produce the fuel. For all fuels, the energy input is in the same fuel.

TABLE 7.1
Fuel Production

Path	Pictorial Description	Energy Ratio (Out/In)
Gas well to user via pipeline		0.98 [1] 0.96 [2] 0.85 [1]
Gas well to user via LNG (liquefied natural gas)		0.9 0.81
Oil well to refinery		0.995 [1] 0.96 [2]
Open-pit coal mine to nearby user		0.99 [1]
Underground coal mine to user		0.98 [2] 0.85 [1]

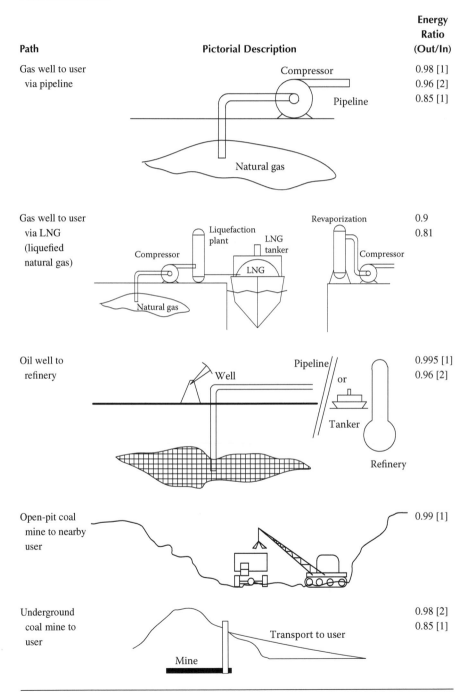

ENERGY CONVERSION/DISTRIBUTION EFFICIENCIES

Table 7.2 lists the technologies and their energy efficiencies for conversion from fuel to electrical energy. Of course, there are ranges of conversion efficiencies for all technologies, functions of capacity, and climatic conditions. However, we include

TABLE 7.2
Fuel to Electricity

Path	Pictorial Representation	Energy Ratio (Out$_e$/In)
Gas turbine, gas fuel to electrical energy		0.38 0.28
Gas turbine/ steam turbine Cogeneration gas fuel to electrical energy		0.60 (Steam-cooled turbine) 0.56
Coal/steam turbine or biomass/steam turbine		0.38 Coal 0.33 Coal 0.28 Biomass [3]
Nuclear, fuel to power		0.33 [4]

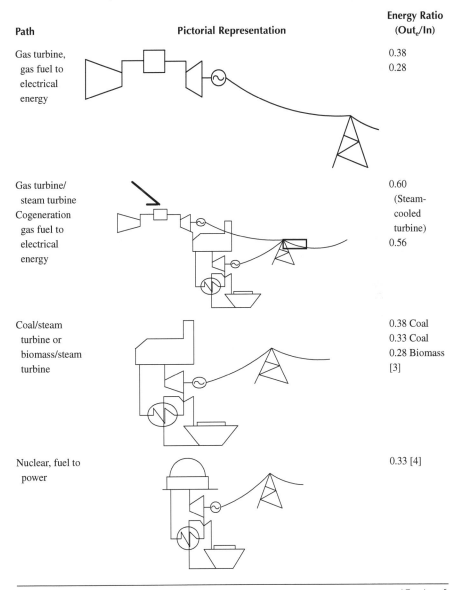

(Continued)

TABLE 7.2 (CONTINUED)
Fuel to Electricity

Path	Pictorial Representation	Energy Ratio (Out$_e$/In)
Natural gas to power, solid oxide fuel cell		0.46 [5]
Natural gas to power, phosphoric acid fuel cell		0.40 0.35 [6]
Gas to power, microturbine		0.33 0.25

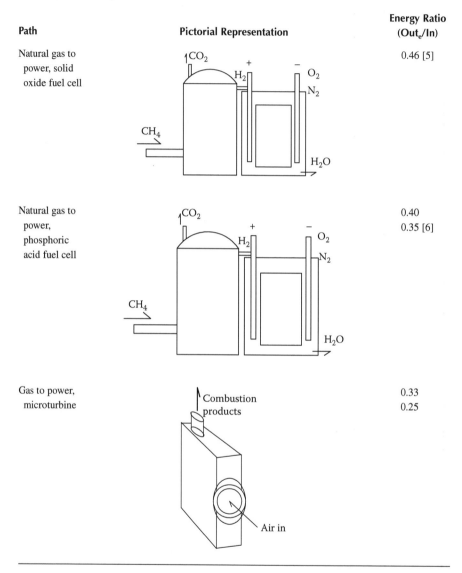

TABLE 7.3
Fuel or Power Distribution

Path	Pictorial Representation	Energy Ratio (Out$_e$/In)
Oil refinery to consumer	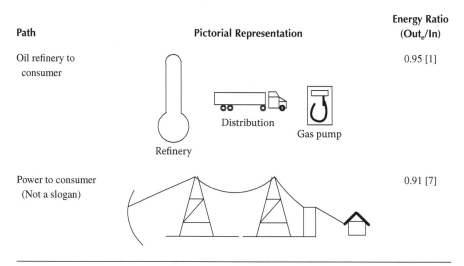	0.95 [1]
Power to consumer (Not a slogan)		0.91 [7]

either typical values or low and high values, on the basis of the lower heating value (LHV) when appropriate.

The distribution of power and fuel takes energy as well, summarized in Table 7.3. Renewable energy falls in the transformation category also (Table 7.4). We list efficiency ranges in terms of their electrical or thermal output to solar/wind input.

STORAGE EFFICIENCY

Storage systems are not in widespread use in the world, except for the universally accepted rechargeable or disposable batteries. However, their widespread adoption could raise peak efficiencies, allowing full capacity operation of energy systems, particularly those based on renewables (Table 7.5).

END-USE EFFICIENCIES

The end-use efficiencies listed in Table 7.6 depend on capacity and frequency of use, and are approximations.

For heating and cooling applications using thermal energy, for small and large capacities, we have the following paths (Table 7.7). Note that residential and commercial equipment serve a residence or an office building, respectively.

TABLE 7.4
Renewables to Heat or Power

Sun to PV power in grid or load		0.24 [8] 0.14 [8] 0.04 [8] organic cells (inexpensive)

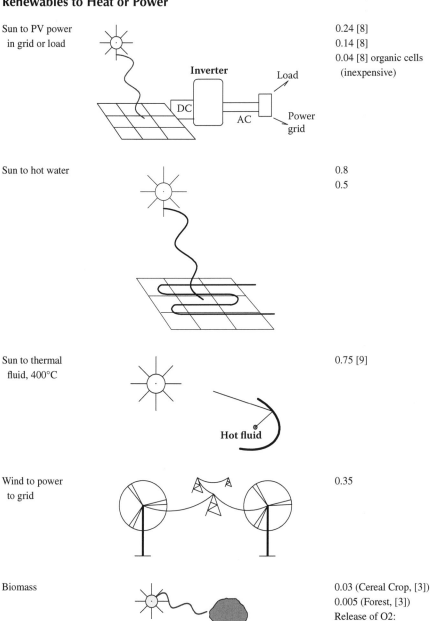

Sun to hot water		0.8 0.5
Sun to thermal fluid, 400°C		0.75 [9]
Wind to power to grid		0.35
Biomass		0.03 (Cereal Crop, [3]) 0.005 (Forest, [3]) Release of O2: priceless

TABLE 7.5
Storage: Electricity, Mechanical, Hydrogen, Thermal

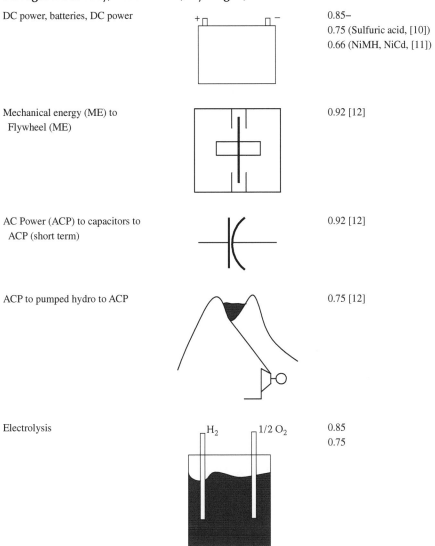

DC power, batteries, DC power	0.85– 0.75 (Sulfuric acid, [10]) 0.66 (NiMH, NiCd, [11])
Mechanical energy (ME) to Flywheel (ME)	0.92 [12]
AC Power (ACP) to capacitors to ACP (short term)	0.92 [12]
ACP to pumped hydro to ACP	0.75 [12]
Electrolysis	0.85 0.75

Electric power can also be used for heating and cooling applications, or to serve electric motors (Table 7.8). The lighting uses are summarized in Table 7.9.

Low overall energy efficiencies are troubling because, for a given amount of energy delivered (or end-use utility), small efficiencies imply larger investment in

TABLE 7.6
End Use: Fuel to Mechanical Power

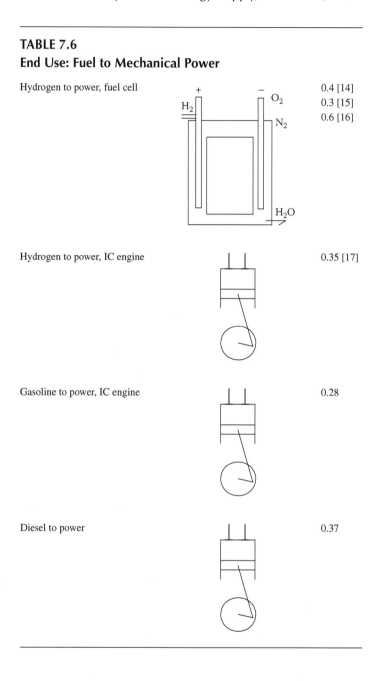

Hydrogen to power, fuel cell		0.4 [14]
		0.3 [15]
		0.6 [16]
Hydrogen to power, IC engine		0.35 [17]
Gasoline to power, IC engine		0.28
Diesel to power		0.37

the upstream chain. Clearly, some energy systems have such low overall efficiencies (see Examples) that one wonders if they could contribute to the energy endeavor. In any case, try your own combinations judiciously, making sure that the inputs and outputs are compatible.

TABLE 7.7
End Use: Fuel to Heat or Cooling

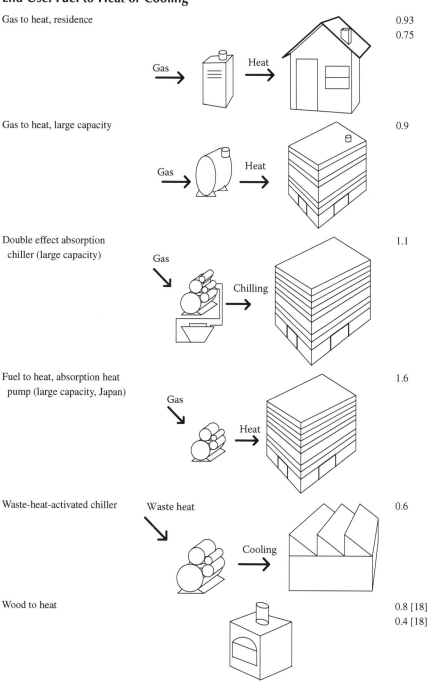

Gas to heat, residence	0.93
	0.75
Gas to heat, large capacity	0.9
Double effect absorption chiller (large capacity)	1.1
Fuel to heat, absorption heat pump (large capacity, Japan)	1.6
Waste-heat-activated chiller	0.6
Wood to heat	0.8 [18]
	0.4 [18]

TABLE 7.8

End Use: Power to Heat or Cooling or Power

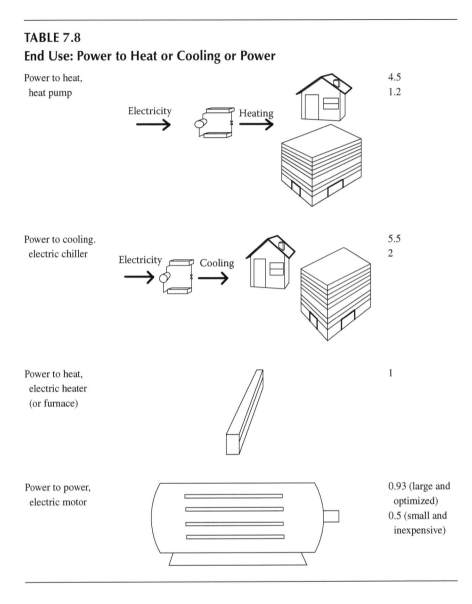

Power to heat, heat pump	4.5 1.2
Power to cooling, electric chiller	5.5 2
Power to heat, electric heater (or furnace)	1
Power to power, electric motor	0.93 (large and optimized) 0.5 (small and inexpensive)

ENERGY ANALYSIS

A crucial if difficult question for a viable transition to other energy sources is that of net energy yield. For instance, if a system requires fossil energy for construction, startup, and operation, then it will be viable only if it will yield more energy than the input during its lifetime. All fossil-fuel-based systems are net energy producers; otherwise, the economies of the 20th century would not have progressed, and they would not exist today. The only known alternatives to fossil fuels are renewables and nuclear. It is relevant then to characterize their net energy

TABLE 7.9
End Use: Power to Light

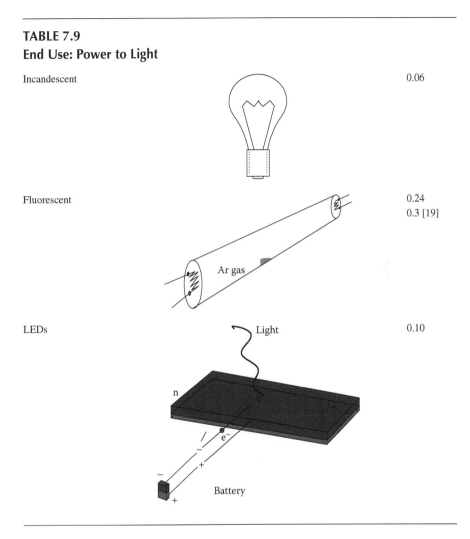

Incandescent		0.06
Fluorescent		0.24
		0.3 [19]
LEDs		0.10

efficiency, given by

$$\eta_{net} = \frac{\text{Energy out over life}}{\text{Fossil energy in}}$$

Even though many uncertainties and discussions concerning these calculations exist, some of the available values are given here. Much discussion takes place around liquid fuels from biomass, in that the net energy efficiency is often reported as less than one or even much greater than one. The numbers recorded (Table 7.10) here apply to efficient processes only, and are deemed feasible.

TABLE 7.10

Net Energy (Over Equipment Lifetime) η_{net}

Sun to PV power in grid or load	27 ([4], amorphous silicon) 8 ([4])
Sun to hot water	83 [20]
Wind to power	51 ([20], 20-yr life) 80 [21]

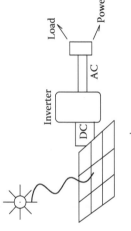

Biomass to fuel

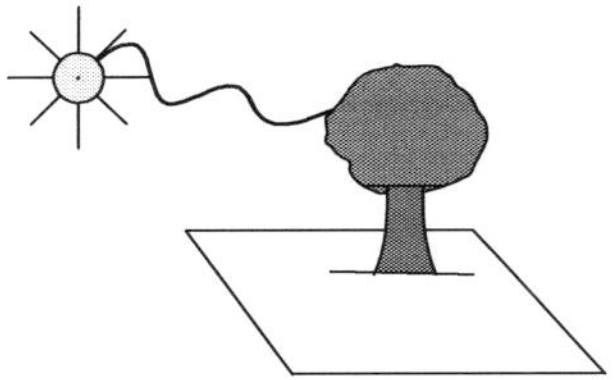

14 [22]
8 [22]

Release of O2: priceless

Solar to ethanol

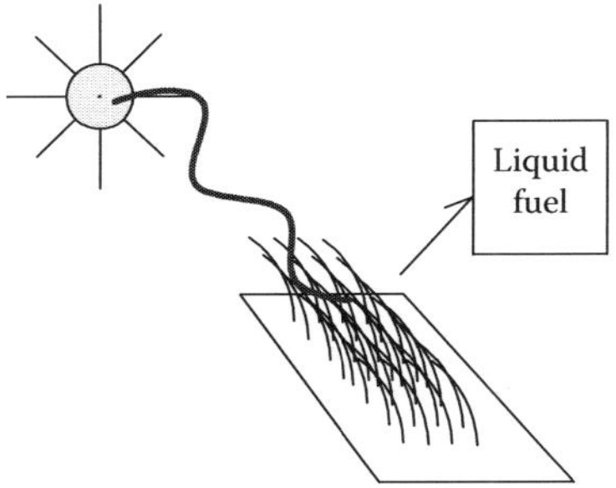

Sugarcane: 8 [23]
Biodiesel: 2.5 [23]
Corn :1.3[a] [3,23]

Liquid
fuel

Nuclear fuel to
electricity

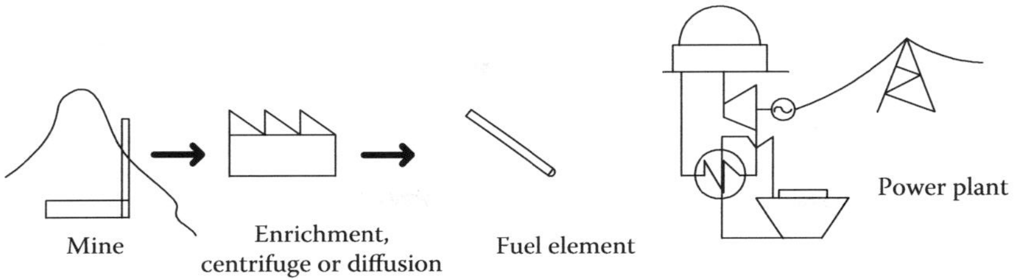

58 (Centrifuge, [4])
17 (Diffusion, [4])
11.6 (One plant, [24])

Power plant

Fuel element

Mine

Enrichment,
centrifuge or diffusion

[a] Energy ratios >1 only if waste is used in production.

TABLE 7.11
Net Energy: Fuel to Fuel

Coal	Divide energy ratio given by Reference 4 (electrical output)	76
	data by 0.38 to obtain thermal input	45
Oil	[25]	20
		8
Gas	Divide energy ratio given by Reference 4 (electrical output)	69
	data by 0.38 to obtain thermal input	
Coal liquefaction[a]	[2]	8.2
		0.5

[a] This value is often quoted as 1, that is, with no net energy yield.

Table 7.11 shows estimates of energy yields of fossil fuels. The net energy yields of Table 7.11 and the efficiencies of previous tables can be combined as necessary for process efficiency evaluation, as shown in Example 7.4.

EXAMPLES

A number of examples accompany this chapter. The reader must be once more advised that these are approximate efficiency values and that only broad conclusions can be extracted from these approximations. The reader may join an energy source and end use with as many processes from the previous tables as his/her imagination may dream of. In Example 7.1 we show two alternative paths to provide heating with natural gas. In locations devoid of temperature extremes, one of the paths appears, from an energy viewpoint, more preferable than the other.

EXAMPLE 7.1 HEATING WITH GAS AND HEAT PUMPS

One may wonder why the residential price of natural gas keeps on increasing. Part of the answer may reside in the two paths considered here. Each starts with gas in the well and finishes delivering one unit of heat to a residence. Whereas regional variations (i.e., outdoor temperature levels) will change the efficiency of Path 2, the path efficiency calculations show a marked difference as to how effectively gas can be used.

Path 1: Well–Processing–Distribution–Combustion in gas furnace and delivery to home (Figure Example 7.1.1)

From Tables 7.1 and 7.7, the maximum and minimum process efficiencies are given in Table Example 7.1.1.

Thus, the maximum expectation is 91%, the minimum is 64%. The second path considered here is

Path 2: Well–Processing–Distribution–GasTurbine Generation–Distribution–Electric Heat Pump (Figure Example 7.1.2)

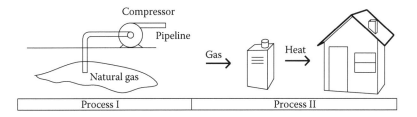

FIGURE EXAMPLE 7.1.1 Path 1 from gas well to residence.

TABLE EXAMPLE 7.1.1
PATH 1 Results

Process I	**Process II**
ηImax = 0.98	ηIImax = 0.93
ηImin = 0.85	ηIImin = 0.75

Overall Efficiencies

ηomax = ηImax × ηIImax = 0.91

ηomin = ηImin × ηIImin = 0.64

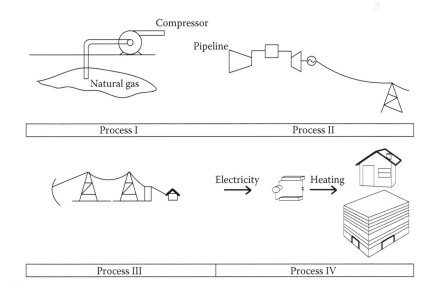

FIGURE EXAMPLE 7.1.2 Path 2 includes transformation into power to heat a house.

TABLE EXAMPLE 7.1.2
PATH 2 Results

Process I	Process II	Process III	Process IV
ηImax = 0.98	ηIImax = 0.38	ηIIImax = 0.91	ηIVmax = 4.5
ηImin = 0.85	ηIImin = 0.28	ηIIImin = 0.91	ηIVmin = 1.2

Overall efficiencies

ηomax = ηImax $\times$ ηIImax $\times$ ηIIImax $\times$ ηIVmax = 1.53

ηomin = ηImin $\times$ ηIImin $\times$ ηIIImin $\times$ ηIVmin = 0.26

We now extract the efficiencies from the tables (7.1, 7.2, 7.3, and 7.8) corresponding to each process (Table Example 7.1.2 PATH 2 results)

Clearly, heat pumps that obtain high efficiencies (such as variable-speed equipment when ambient temperatures are above ~5°C) will perform very well, and Path 2 is then preferable to Path 1. At low ambient, the minimum efficiency of Path 1 exceeds that of Path 2. Hence, in areas with low temperatures for much of the winter, either a ground-coupled or geothermal heat pump must be used to obtain high efficiencies, or a simple gas furnace will result in smaller gas demand.

Cogeneration plants, delivering much better efficiencies than a simple gas turbine, may increase the efficiencies considerably.

Example 7.2 shows that when it boils down to the remarkable fuel that natural gas is, cogeneration technology may be hard to supersede on an efficiency basis alone.

EXAMPLE 7.2 USING GAS TO PRODUCE ELECTRICAL ENERGY

We consider two paths for natural gas utilization. Electricity is produced and shipped to heat homes with heat pumps. In Path 1, we use a reformer-fuel cell to activate the heat pump (Figure Example 7.2.1).

The efficiencies and results for Path 1 are shown in Table Example 7.2.1 (from Tables 7.1, 7.2, 7.3, and 7.8, identifiable with the same diagrams of Figure Example 7.2.1).

TABLE EXAMPLE 7.2.1
PATH 1 Efficiency and Results

Process I	Process II	Process III	Process IV
ηImax = 0.98	ηIImax = 0.46	ηIIImax = 0.91	ηIVmax = 4.5
ηImin = 0.85	ηIImin = 0.46	ηIIImin = 0.91	ηIVmin = 1.2

Overall efficiencies

ηomax = ηImax $\times$ ηIImax $\times$ ηIIImax $\times$ ηIVmax = 1.85

ηomin = ηImin $\times$ ηIImin $\times$ ηIIImin $\times$ ηIVmin = 0.43

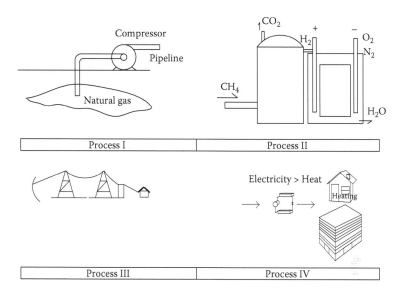

FIGURE EXAMPLE 7.2.1 Path 1: Gas well–SOFC–heat pump.

In the second path, the SOFC is replaced by a large cogeneration plant to generate the power (Figure Example 7.2.2). The efficiencies in Table Example 7.2.2 are extracted from Tables 7.1, 7.2, 7.3, and 7.8.

It is apparent that Path 2, involving the cogeneration plant, has today higher efficiencies than the SOFC. The possibility of using the waste heat from either cycle

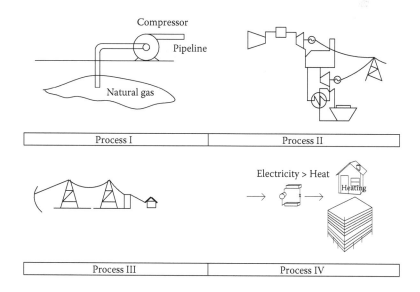

FIGURE EXAMPLE 7.2.2 Path 2: Gas well–Cogeneration Plant–heat pump.

TABLE EXAMPLE 7.2.2
PATH 2 Efficiency and Results

Process I	Process II	Process III	Process IV
ηImax = 0.98	ηIImax = 0.6	ηIIImax = 0.91	ηIVmax = 4.5
ηImin = 0.85	ηIImin = 0.56	ηIIImin = 0.91	ηIVmin = 1.2

Overall efficiencies

ηomax= ηImax × ηIImax × ηIIImax × ηIVmax = 2.4

ηomin= ηImin × ηIImin × ηIIImin × ηIVmin = 0.52

for direct heating is certainly a consideration, but only when the power and thermal demands coincide.

In Example 7.3, we look at coal and transportation, but only at two paths. The user is encouraged to find other alternative paths (even from his/her own research).

EXAMPLE 7.3 RIDING ON WATER AND COAL: AT WHAT ENERGY COST?

The United States is richly endowed with coal, as seen in Chapter 3. Hence, why not use it to produce hydrogen? To contribute to this answer, it is appropriate to compare a new technology (fuel cell) to a well-known technology (electric motors in trains, subways, etc.). One could argue that the destination of both paths is not the same, for one is geared to individual transportation, the other to mass transportation. We argue that the end use is transportation and define the two paths. For Path 1, we have five processes, from mine to power to electrolysis to fuel cell. (Other means of obtaining the hydrogen by thermal breakdown are possible, but we project that distributing power is simpler than distributing hydrogen.) (Figure Example 7.3.1.)

As shown in Table Example 7.3.1, with efficiencies from Tables 7.1, 7.2, 7.3, 7.5, and 7.6 (identify cases with the same schematic), the efficiencies of Path 1 are quite low. Also, clean water would be required.

TABLE EXAMPLE 7.3.1
PATH 1 Efficiencies and Results

Process I	Process II	Process III	Process IV	Process IV
ηImax = 0.98	ηIImax = 0.38	ηIIImax = 0.91	ηIVmax = 0.85	ηIVmax = 0.4
ηImin = 0.85	ηIImin = 0.33	ηIIImin = 0.91	ηIVmin = 0.75	ηIVmin = 0.3

Overall efficiencies

ηomax = ηImax × ηIImax × ηIIImax × ηIVmax × ηVmax = 0.11

ηomin = ηImin × ηIImin × ηIIImin × ηIVmin × ηVmin = 0.06

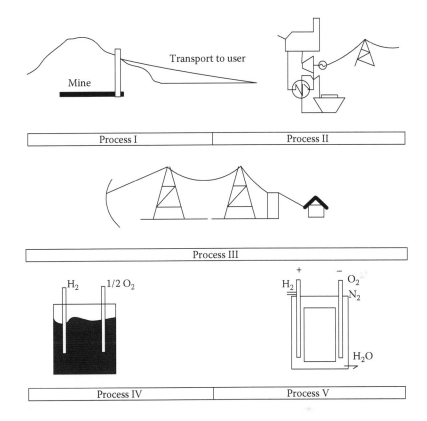

FIGURE EXAMPLE 7.3.1 Path 1 goes from coal to electrolysis, using current power generation efficiencies.

Let us compare the results of Path 1 to those of Path 2, namely, the production of electric power for a third rail with current efficiencies. We show the path in Figure Example 7.3.2, and the efficiencies from Tables 7.1, 7.2, 7.3, and 7.8.

The efficiencies of Path 2 are larger, as shown in Table Example 7.3.2.

TABLE EXAMPLE 7.3.2
PATH 2 Efficiencies and Results

Process I	Process II	Process III	Process IV
$\eta Imax = 0.98$	$\eta IImax = 0.38$	$\eta IIImax = 0.91$	$\eta IVmax = 0.93$
$\eta Imin = 0.85$	$\eta IImin = 0.33$	$\eta IIImin = 0.91$	$\eta IVmin = 0.5$

Overall efficiencies

$\eta omax = \eta Imax \times \eta IImax \times \eta IIImax \times \eta IVmax = 0.31$

$\eta omin = \eta Imin \times \eta IImin \times \eta IIImin \times \eta IVmin = 0.14$

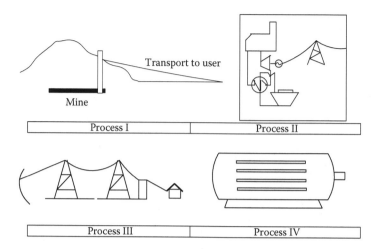

FIGURE EXAMPLE 7.3.2 Path 2 goes from coal to power, using current power generation efficiencies.

If the goal of any commercial venture is to provide a service to as many as possible with as high an efficiency as possible, clearly the "third rail" (Path 2) option may offer economic enticements that the technology of Path 1 could not.

The topic of net energy yield is taken up in the last example. The point to be made is that, renewables, due to their low energy density, may need to be used in conjunction with energy conservation technology to succeed.

EXAMPLE 7.4 A SUSTAINABLE TRANSITION?

The United States is richly endowed with solar. What would be the perspective of using solar versus gas for power generation and heat pumps? Would a sustainable transition be possible if all the difficulties associated with solar were met with superior engineering?

Consider Path 1, from solar to heat pump (Figure Example 7.4.1). The net energy yield is projected with figures from Tables 7.10, 7.3, and 7.8 (Table Example 7.4.1).

TABLE EXAMPLE 7.4.1
Yields for Path 1

Process I	Process II	Process III
ηInetmax = 27	ηIImax = 0.91	ηIIImax = 4.5
ηInetmin = 8	ηIImin = 0.91	ηIIImin = 1.2

Overall efficiencies

ηomax = ηInetmax × ηIImax × ηIIImax = 110

ηomin = ηInetmin × ηIImin × ηIIImin × = 8.7

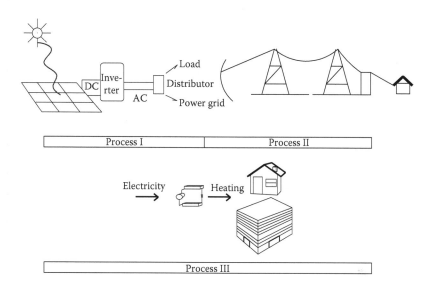

FIGURE EXAMPLE 7.4.1 Path 1: from solar to heat pump.

The reader may note that we are listing the energy yield for the energy-supplying technology (solar in this case), and the efficiency for all the others. This means that we assume the other processes to be already in place, or that they will be implemented no matter what the source might be. If solar is simply used to feed power to the grid, the adopted outlook should hold. The results are shown in Table Example 7.4.1. The table shows that if all solar energy were used at the efficiencies of processes II and III for the equipment lifetime, a transition (from a sustainability viewpoint) to solar energy would be possible. However, the net energy yield of other technologies is not small either, and shows that economics and energy ratios may be more strongly linked than one would surmise.

Consider instead the conversion gas to power to heat pump in terms of energy yield, Path 2 (Figure Example 7.4.2). The energy yields are extracted from Tables 7.1, 7.2, 7.3, and 7.8. The numbers are recorded in Table Example 7.4.2.

TABLE EXAMPLE 7.4.2
Figures for Path 2

Process I	Process II	Process III	Process IV
ηInetmax = 69	ηIImax = 0.6	ηIIImax = 0.91	ηIVmax = 4.5
ηInetmin = 69	ηIImin = 0.56	ηIIImin = 0.91	ηIVmin = 1.2

Overall efficiencies

ηomax = ηInetmax × ηIImax × ηIIImax × ηIVmax = 169

ηomin = ηInetmin × ηIImin × ηIIImin × ηIVmin = 42.2

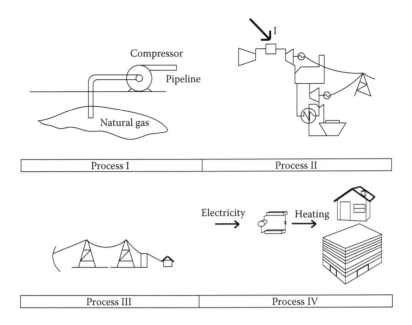

FIGURE EXAMPLE 7.4.2 Path 2: from gas to heat pump.

Inspecting the results, one can say that the energy payback for natural gas is quite large with high-end technology. Hence, investment in advanced, proven technology that favors energy conservation stretches gas supplies and also enhances the viability of solar energy. Either technology can sustain a transition, but solar seems more abundant than natural gas, although it does require back-up or storage.

REFERENCES

1. Smil, V., 2003, *Energy at the Crossroads*, MIT Press, Cambridge, MA.
2. Cleveland, C. J., Constanza, R., Hall, C., Kaufmann, R., 1984, Energy and the US economy: A biophysical perspective, *Science*, New Series, Vol. 225, No. 4665, 1984, pp. 890–897.
3. Twidell J., Weir, T., 2005, *Renewable Energy Resources*, Taylor and Francis, London.
4. World Nuclear Association, 2006, http://world-nuclear.org/info/inf11.htm
5. Siemens, 2005, Solid Oxide Fuel Cells, Brochure from "Stationary Fuel Cells," 1310 Beulah Rd., Pittsburgh, PA 15235-5098, www.powergeneration.siemens.com/en/fuelcells
6. California Energy Commission, http://www.energy.ca.gov/distgen/equipment/fuel_cells/performance.html
7. Energy Information Agency, http://www.eia.doe.gov/emeu/aer/pdf/pages/sec8_3.pdf
8. National Renewable Energy Laboratory, http://www.nrel.gov/ncpv/thin_film/docs/1
9. Quaschning, V., Solar Thermal Power Plants: Technology Fundamentals, *Renewable Energy World* 06/2003, pp. 109–113, http://www.volker-quaschning.de/articles/fundamentals2/index_e.html

10. Transtronics, 2006, http://xtronics.com/reference/batterap.htm
11. Powerstream http://www.powerstream.com/NiMH.htm
12. Climate change, *U.S. Climate Change Technology Program—Technology Options for the Near and Long Term,* August 2005, pp. 1.3–11.
13. Mac Queen, B., http://gcep.stanford.edu/pdfs/hydrogen_workshop/MacQueen.pdf
14. Crea, D., 2005 Fuel Cell Efficiency: A Reality Check, http://www.evworld.com/view.cfm?section=article&storyid=730
15. U.S. DOE, 2003 http://www.eere.energy.gov/hydrogenandfuelcells/pdfs/gronich_garbak_ee_tech_val.pdf
16. U.S. DOE, 2003 (1), Fuel Cell Report to Congress, http://www.eere.energy.gov/hydrogenandfuelcells/pdfs/fc_report_congress_feb2003.pdf
17. PFL, http://planetforlife.com/h2/h2vehicle.html
18. EPA, 2006, http://www.epa.gov/woodstoves/efficiently.html
19. Forcefield Group, 2008, http://www.otherpower.com/otherpower_lighting.html
20. Tiwari G. N., Ghosal, M. K., 2003, *Renewable Energy Resources,* Alpha Science International. Harrow, U.K.
21. Danish Wind Industry Association, http://www.windopwer.org/composite-53.htm
22. Nussbaumer T., Oser, M., 2004. Evaluation of Biomass Combustion-Based Energy Systems by Cumulative Energy Demand and Energy Yield Coefficient, IEA report. Also in http://www.ieabcc.nl/publications/Nussbaumer_IEA_CED_V11.pdf
23. Bourne, J. K., 2007, Green Dreams, *National Geographic,* Oct. 2007, Vol. 212, No. 4, pp. 38–59.
24. Tyner, G., 2002, Net Energy from Nuclear Power, http://www.mnforsustain.org/nukpwr_tyner_g_net_energy_from_nuclear_power.htm
25. Cobb, K. 2006. "Should we use net energy to measure global energy reserves?" http://resourceinsights.blogspot.com/2006/04/should-we-use-net-energy-to-measure.html

8 Energy and Its Sequels

… when the possession of Earthly things is under discussion, men find it difficult to reason with justice…

Umberto Eco, in *The Name of the Rose*

As energy is used or harvested, waste heat and various pollutants, over which no one claims ownership, are produced. We broadly discuss these products, their origin, and their potential effects on the climate and on humans. The material of Chapter 3 is used to speculate on the time scale of abating some of the pollutants.

CONSTANT TRAVEL WITH A BRIEF STOP ON EARTH

As ascertained by Albert Einstein, matter and energy are somewhat equivalent [1]. We say "somewhat," for the conversion of matter into energy, which is difficult, seems much easier than the reverse. Let us look at energy present as matter in the Sun, released by fusion, and destined to visit, if only temporarily, this planet. We now avail ourselves of the liberties of science fiction. Let us imagine that we are a J in transit. Then, some time ago, we were traveling from the Sun to the Earth (at the speed of light, no less). We then were absorbed, became part of a living organism that died and whose corpse decayed, and over eons and through different processes, a fraction of us became coal or oil or natural gas. On the other hand, if this did not happen eons ago, then we were absorbed into the atmosphere, causing wind, which was converted to power by a windmill, or simply absorbed by a photocell to become power, or by a solar collector to become heat. Although we would not know where we came from (we use star and condensed matter as explanations), we could also have been in $^{235}U_{92}$ and survived an unimaginable period of time, seeing half of the host atoms disintegrate every 700 million years (i.e., like living in a hotel that is being demolished systematically). Yet another possibility is that we originated from friction and atomic decay in the core of the planet to be labeled geothermal energy. In any case, today we were released from our bond/atomic binding force/pressure difference or whatever, becoming energy in transit from our previous (comfortable) existence as stored energy.

As we were released, we became power or heat flow. If we became power, we saw two-thirds of our J mates go to waste heat. Then, if we were used for heating, cooling, or lighting, we ended up in the atmosphere. Perhaps we ended up lifting a beam in a skyscraper, and now we are potential energy for a short (from a geological viewpoint) while. If we became heat flow, we ended up in some chemical bonds (plastic, food, paper) or in the atmosphere as well, as heat was lost from a stove to the house to the outdoors. If we became embodied in a fuel, we moved something (plane or car). If those inventions returned to the same elevation (and most of them do, on average), we ended up as heat in the atmosphere. In flames or atomic reactors, some by-products of dubious utility are formed. They also end up in the atmosphere, or they produce heat that ultimately ends up in the atmosphere.

So, our trip continues: we can be temporarily stuck as either potential or chemical energy, but once released, our thirst for travel is unquenchable. We are, however, limited as to our destinations: wanderlust ultimately takes us to lower temperatures or to lower chemical or gravitational potentials. However, there is more: the processes we underwent to render utility release a number of chemical compounds that also end up in the environment that we inhabit. These multiple transformations were appraised, if only dimly, in Chapter 7.

We summarize these science fiction paragraphs in Figure 8.1. Everywhere the agency of time is indispensable, from minutes to eons in the case of the figure. The formation of fossil fuels requires time, and so does the conversion of nuclear waste to waste heat. Only renewables are on a "pay as you go" basis. The Second Law always wins: some of the energy ends up as heat no matter what. This is another way of saying that Carnot [2] was right from the beginning of the saga of energy transformation. It is, however, significant that renewables produce electricity (if not cheaply) in enough quantity to produce their own harvesting technology, as seen in Chapter 7. Our trip does not end here: as the atmosphere slowly gets rid of heat, we end up (possibly) in space as a very, very cold J (about 3 K).

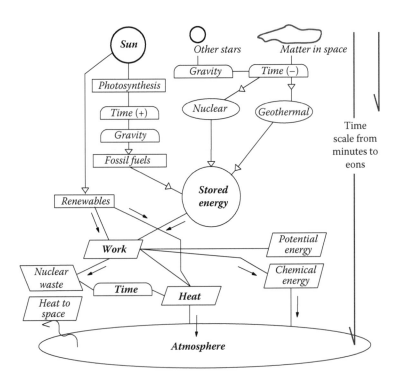

FIGURE 8.1 The flow of energy in time.

There is more: the Universe is expanding, we presumably still travel, but this is hard to determine. We now abandon the unconfined comfort of science fiction, we hope.

What is then the point? The point is to illustrate the Second Law: the J of our story, which is that of all Js, flowed from potential energy in motion to the "null" (in terms of utility) potential of the coolest temperature available. The descent was predicted, and, as long as physical laws remain unchanged, unavoidable. This descent generated matter and waste heat because there does not seem to be a way around it. In this chapter, we cover what ends up in the atmosphere or environment for a sufficiently long time as to potentially affect the activities of humans.

COMBUSTION

We convert a lot of this planet's energy via combustion. The products no one wants to claim, as the quote at the beginning of this chapter suggests. Combustion produces gases and products that linger in the atmosphere for a long time. None of them can be called beneficial, except that an abundance of CO_2 might perhaps favor plant growth, although the evidence is not clear in that respect. In any case, the products depend largely on the fuel under consideration and on the way its combustion proceeds. Crucial parameters for describing, if only briefly, what may happen in a combustion process are

    a. The equivalence ratio
    b. The local flame temperature

Every fuel needs a certain amount of air for complete combustion. The proportion of fuel to air is called the stoichiometric fuel-to-air ratio (in mass, typically). When a fuel is burned, the fuel-to-air ratio does not necessarily match the stoichiometric one. Hence, the equivalence ratio, defined as

$$\Phi = \frac{fuel\text{-}to\text{-}air\ ratio}{fuel\text{-}to\text{-}air\ ratio\ stoichiometric}$$

measures the departure from the stoichiometric ratio. The local flame temperature is somewhat related to the local value of Φ. More importantly, the local flame temperature largely determines what compounds will form during combustion.

NATURAL GAS

This fuel is regarded as having comparatively small impact on the environment. Its combustion generates CO_2 in the smallest amount per unit of delivered thermal energy. Also, some oxides of nitrogen (NO_x) are typically generated if the flame temperature is high, which entails an equivalence ratio of 1. (See Figure 8.2, typical of gas turbine combustors.) An effort is generally made to achieve "lean" combustion [3] (fuel/air ratio below stoichiometric or equivalence ratio smaller than one) because lower flame temperatures decrease the amount of NO_x. Even leaner

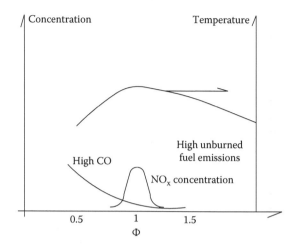

FIGURE 8.2 Pollutants and flame temperature for different equivalence ratios.

combustion decreases the NO_x further, but it increases the carbon monoxide (CO) concentration as the temperature decreases and the activation energy barrier cannot be overcome. Conversely, "rich" combustion (with excess fuel instead of excess air) generates unburned fuel, reducing efficiency and releasing noxious fuel. Some natural gases may possess sulfur compounds that result in SO_2 production, leading ultimately to acid rain.

OIL

Oil is regarded as having a large impact on the environment, simply because of its use in IC engines. These engines have a highly transient combustion process, as opposed to that of gas turbines, which is a fairly smooth, steady-state process by comparison. As shown in Example 8.1, the time available for combustion, even when the engine turns at low speeds, is quite brief.

EXAMPLE 8.1 THE IC ENGINE AND THE FREQUENCY OF COMBUSTION

There are about 500 million cars in the world today (Chapter 4). Although we do not know the average number of cylinders or how many are operating simultaneously, let us assume that, at any given time, one-fifth of the cars are operating on a minimum of 4 cylinders at 2500 rpm.

With the given data, we estimate the number of explosions per second and their duration, using some data from the literature.

$$\text{Number of cars} \qquad noc = 500 \cdot 10^6$$

$$\text{Turns per minute} \qquad rpm = 2500 \, \frac{1}{\text{min}}$$

As seen in Chapter 4, there is an explosion in each cylinder every two turns of the engine. Hence,

Explosions per cylinder $\quad excyl = \dfrac{rpm}{2} = \dfrac{2500}{2} \cdot \dfrac{1}{60 \cdot \text{min}} = 20.8\,\dfrac{1}{\text{s}}$

Because our engines have four cylinders and 20% are operating, we have

Explosions per second globally $\quad W\exp\sec = excyl \cdot 4 \cdot 0.2 \cdot noc = 8.3 \cdot 10^9\,\dfrac{1}{\text{s}}$

This many explosions per second is hard to imagine, but it explains the voracity for oil, and the pollution, and the profits of spark-plug manufacturers. Let us calculate the duration of each explosion. According to Reference 4, an explosion releases the most energy during the time interval corresponding to 10° of crank angle. Adopting this value, the fraction of one turn devoted to the explosion is

Time fraction $\quad tf = \dfrac{10}{360} = 0.028$

The duration of one turn at 2500 rpm is

$$tturn = \dfrac{1}{rpm} = \dfrac{60\,\text{s}}{2500} = 0.024\,\text{s}$$

Hence, the lifetime of an explosion for the assumed conditions is brief:

$$t\exp = tturn \cdot tf = 0.024 \cdot 0.028\,\text{s} = 0.00067\,\text{s}$$

Since each explosion contributes its quota of pollutants, it is clear that the abundance of IC engines makes managing their combined pollution quite a challenge.

Predicting or measuring the temperature of the gas in the cylinder is a major undertaking. Based on the results of Reference 4, a qualitative temperature signature versus crank angle (0 angle denotes beginning of intake) is shown in Figure 8.3. This signature is helpful to explain how the reaction between nitrogen (N_2) and oxygen (O_2) to form nitrogen oxides (NO_x) comes about. According to the combustion models [3], most of the NO_x forms when the temperature extremes take place, that is, when the gas is confined to a small volume and the temperature is high. Those conditions ensure a large number of molecular collisions. As the gas expands in the cylinder, its temperature drops, which would favor the recombination of N_2 and O_2. However, with increasing volumes and reduced temperatures, there is not enough time (i.e., collisions) for the NO_x to revert to N_2 and O_2. These engines, due to the high temperatures and rapid gas cooling associated with in-cylinder combustion (see Chapter 4), tend to produce large amounts of NO_x.

CO tends to arise in cylinders running rich, that is, with an equivalence ratio Φ greater than one. There is not enough O_2 to oxidize the fuel completely, and some CO remains. This is behavior opposite to that of the gas turbines (Figure 8.2). Unburned hydrocarbons are more common with oil products than with natural gas because the

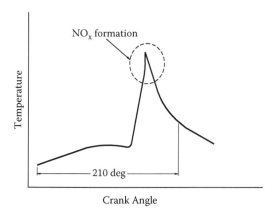

FIGURE 8.3 Sketch of in-cylinder temperature profile.

complete combustion of liquid fuels can present challenges: vaporizing liquid droplets may not always be complete, and hence combustion will not be. Soot can then be present.

The exhaust of the engines needs to be cleaned via Pt catalysis to decrease the NO_x content. Impurities in oil result in an assortment of noncombustibles called ash and particulates [5]. Sulfur compounds can also appear.

It is interesting to quote some figures indicating the extent of pollution by different IC engines [6]. For subcompact cars, the CO emitted is 0.3 g/mi, and the NO_x is 0.04 g/mi. For a compact car, values are 1.3 g/mi for CO and 0.2 g/mi for NO_x. Hybrid cars achieve much smaller emission levels. When considering the life cycle of the fuel, the differences are not so pronounced, but the interested reader is advised to look more deeply into the matter.

COAL

Now here is a difficult fuel. Coal is not easy to extract, produces the most CO_2 per unit energy released, can have SO_2, NO_x, mercury, and the ashes, and particulates can contain uranium and thorium oxides [7,8,9], just to mention a few possibilities. Because nearly 41% of the worldwide and 51% of the U.S. power generation is coal based, this fuel is of evident utility but will have a considerable effect on the environment.

The effect of CO_2 on climate is a topic of discussion, but suffice it to say that per unit of energy delivered by combustion, coal produces the most CO_2. For instance, it has been estimated [10] that per unit of energy (same quality of energy, that is) coal produces 73% more CO_2 than gas, and 18% more than oil. Because of the nature of coal deposits, mercury is often present, and whatever is not removed is released to the environment, where it can be absorbed by living organisms with adverse effects. Most plants have gas treatment equipment which we describe in the following text. A recent study [7] concluded that power plants with electrostatic precipitators and desulfurization equipment can remove up to 75% of the mercury. If the plant is equipped with selective catalytic reduction for NO_x removal, then up to 90% of the mercury can be removed.

Sulfur oxidizes to sulfur dioxide that, when released into the atmosphere and combined with water and oxygen, produces acidic compounds that result in acid rain. Sulfur oxides are typically removed in wet scrubbers, where a basic solution (hydrated lime) is sprayed on the gases. The product of this process is gypsum. There are other technologies to remove sulfur compounds as well.

Particulates can be removed by electrostatic precipitation (ESP) or filtration in bag-houses or wet scrubbing [7]. In an electrostatic precipitator, an electric field acts on the charged particles to impose a trajectory that results in collision with a solid surface and removal from the flow. The efficiency of ESP can be very high, on the order of 99.8% removal. Baghouses and wet scrubbing can also reach high collection efficiencies.

The NO_x can be removed by selective catalytic reaction (SCR) or selective noncata-lytic reaction (SNCR) [7]. In the former process, ammonia is injected into the gas stream and reacts with the NO_x in the presence of a catalyst, typically TiO_2 [11]. In the SNCR, no catalyst is present, and the reduction occurs by a number of reducing chemicals.

Coal can be transformed into syngas (synthesis gas) of varying H_2 content in gasifiers, downstream of which the gas can be cleaned off and combusted in a high-efficiency cycle. A process schematic is shown in Figure 8.4. The gasifier is not

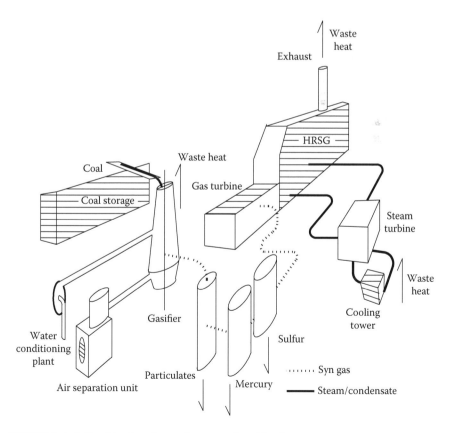

FIGURE 8.4 A Combined cycle running on coal gasification.

unlike the blast furnace of Chapter 6, in that the coal is fed at the top and it reacts with steam and oxygen as it flows downwards, exiting as syngas and ash. Naturally, feeding a gasifier operating at high pressure is not an easy technical problem, and neither is it easy to clean and extract pollutants and particulates. In any case, multiple challenges must be overcome on a continual basis for the concept to be viable.

Even when coal is gasified, the process is such that the gas will not damage gas turbines but will generate ashes and other compounds all the same. In any case, coal accounts for most of the electricity production in the world, and the combined cycles possible with gasification have high thermal efficiencies. Hence, much can be expected from progress in the gasification area. More will be said about the future, that imponderable time domain, in Chapter 10.

ALL TOGETHER NOW

Today, the main concern surrounding fossil fuels hinges on the production of CO_2 and its unintended consequence: global warming. Like water vapor, CO_2 absorbs radiant energy in bands (infrared) that result in thermal energy staying on Earth, instead of being rejected into space. Hence, more CO_2 in the atmosphere means that more heat will be trapped on this planet. In the absence of compensating mechanisms, such as increased reflectance, the current atmospheric temperature will increase. The evidence for global warming is hotly debated.

Please note that the author did not write debatable, the author wrote debated. The basics of the problem are straightforward: if enough infrared absorbing gases are released, their combined effect will be to trap radiant energy so as to require some other mechanism to intensify heat rejection if the temperature of this world is to stay, on the average, the same. Of course, the other solution is to stop ejecting CO_2, which, although a solution, poses all kinds of other problems. Identifying a mechanism or mechanisms to discount or amplify the effect of CO_2 has demanded much intellectual energy, but the search, so far, has been fruitless in stopping CO_2 buildup or, as many maintain, its consequences.

What is certain is that, as the concentration of CO_2 has increased, so has the surface temperature of the planet [12]. According to reference 12, average surface temperatures have increased since 1950 by 0.5 K, while ocean level has risen by 5 cm, and snow cover in the Northern hemisphere has decreased by $2 \cdot 10^6$ km^2. So many other things have also changed, such as stock markets, the rate of inflation of some currencies, the nominal price of copper, and the number of people on the planet with and without cars, etc. In other words, critics of this theory claim that the evidence is sketchy. Others firmly believe in it. The magnitude and speed of the phenomena are so much beyond the scale of a generation that many choose to ignore it and try to enjoy life without concerns.

Denial, when it comes to the planet one inhabits, is not helpful or advisable. Historically, things have been better in some places than others depending on geography and those factors that we call economics, politics, and fate. Global warming will affect all [12] indiscriminately, but its consequences will be, as everything that concerns nature, distributed unevenly in space and time. Hence, the two opposing views must be reconciled, and a coherent action plan must evolve for the sake of the quality of living on this planet.

This author thinks that global warming is no longer debatable. However, there are many who continue to debate, if not the fact, the causes. The main difficulty in reconciling these views is that accepting "global warming" as a manifestation of CO_2 emissions implies that current lifestyles must change, and inertia is not just a word but a way of life. It is an established trait of the human species (on all levels) to expect that others will modify their conduct. However, if climate change is to be mitigated, individual choices of supplier and end-use technology could make a considerable difference.

In any case, less CO_2 implies one of three things or a combination of all of them: a life with less energy, a transition to large-scale nuclear power while having to deal with the ensuing waste, or a transition to the still unproven (at least in massive scale) renewable technologies. Whereas some groups are convinced and convincingly argue about the benefits of one of those futures, most folk just desire the energy; but few seem to want to face any of those futures squarely. In theory at least, what is economically viable and results in greater benefit for generations to come should prevail. In practice, and because it is hard to determine what the benefits are beyond, perhaps a generation, the future is hard to predict, and the outcome will be a mixture of many interests and ideals.

Concerning the other pollutants, NO_x and SO_2 result in acid rain. NO_x can form smog, and in general contributes a lackluster brown tint to the atmosphere. It can irritate the respiratory system of humans and react with other compounds to produce toxic products. SO_2 has been associated with increased death rates due to cancer and cardiopulmonary disease [13].

NUCLEAR

Nuclear fission, as explained in Chapter 5, produces energy, waste heat, and spent fuel, but other than the fossil fuel used for construction and operation, produces no CO_2. According to the World Nuclear Association [14], the spent fuel is mostly $^{238}U_{92}$ (~95%), 1% $^{235}U_{92}$, 3% stable fission products, and about 1% plutonium. Of this 1%, about two-thirds are $^{239}Pu_{94}$ and $^{241}Pu_{94}$, both fissile, and the rest is $^{240}Pu_{94}$, which is not that easily split. Separation of the high-level waste (HLW; waste that must be cooled and shielded for some time to avoid damage) results in a mixture that in about 9000 years returns to the same level of radioactivity as uranium ore. The low-level waste (LLW; not very active, contributing only 1% of the radioactivity of all nuclear waste) is encased in concrete and buried in shallow pits. Intermediate-level waste (ILW), mostly from decommissioned reactors or parts of them, is disposed of as low-level waste or as high-level waste, depending on its active life.

Disposal of HLW requires planning. As indicated, it requires shielding to reduce the radiation ensuing from it, and cooling. There is no easy way to dispose of HLW in the United States, where it currently lays on pools or dry casket storage in nuclear plants. The plan for long-term storage is to develop a repository in Yucca Mountain, Nevada, inside canisters with a 10,000-year life. This repository has a projected capacity of 70,000 tonnes of waste. In Example 8.2, we discuss the available capacity as related to the waste needing storage.

EXAMPLE 8.2 COMPARISON OF WASTE GENERATED AND STORAGE CAPACITY

We seek to obtain a ratio of waste generated to mined uranium, the "yellow cake" of Chapter 4. With this ratio and the knowledge of U reserves, we can project how much waste could be generated with current technology. According to the World Nuclear Association [14,15], a 1,000-MWe plant generates 27 tonnes of waste in one year. Also, yearly nuclear power production of 370 GWe requires 67,000 tonne/yr of mined U oxide [15], plus the equivalent of 10,600 tonne of mined U but extracted from decommissioned nuclear weapons instead. Hence, one plant of 1,000 MWe would require (on the average)

$$Mined\ U = \frac{(67,000 + 10,600) \cdot \dfrac{tonne}{yr}}{370 \cdot GWe} \cdot 1,000\ MWe \approx 209\ \frac{tonne}{yr}$$

Since the plant generates 27 tonne/yr, we have a ratio of waste to mined uranium of

$$Ratio = \frac{27}{209} \approx 0.13$$

Assuming this ratio to be correct, let us project the required storage needs for existing U reserves. In Chapter 3, we gave the values for uranium reserves in the United States as

$RUSA_{url} = 1.20 \cdot 10^5$ tonne at \$66,000/tonne
$RUSA_{urh} = 4.04 \cdot 10^5$ tonne at \$110,000/tonne

So, multiplying the reserves by the 0.13 factor, we obtain

~15600 tons of waste, at \$66,000/tonne, or
~52000 tons of waste, at \$110,000/tonne

This waste would be stored in Yucca Mountain. The repository has a projected capacity of about 70,000 tonne of waste, and there are already 49,000 tonne of waste distributed in the country, plus 12,000 tonne from weapons programs. Thus, an uncommitted capacity of 9,000 tonne [19] seems to exist. Comparing the available 9,000 tonne capacity to the waste to be produced as the U reserves are consumed, we have for inexpensive and expensive reserves

$$Inexpensive\ factor = \frac{9000}{15600} \approx 0.6$$

$$Expensive\ factor = \frac{9000}{52000} \approx 0.2$$

Currently, the Yucca Mountain repository could supply between 0.2 and 0.6 of the capacity required by today's reactors to use the reserves within the United States. Reprocessing could alleviate the storage problem for a number of years, but it will not eliminate it. Once a waste is generated, a suitable place for long-term storage must be found. Unfortunately, such a place is, with current technology, nowhere but on this Earth.

The fuel can be reprocessed, and this is done in France and England, and planned in Japan. The spent fuel is treated to separate the remaining $^{235}U_{92}$ and the produced $^{239}Pu_{94}$. Those atoms are used to prepare new fuel, called mixed oxide fuel (MOX), which can then power reactors designed to this effect. According to the World Nuclear Association [15], the Pu in the MOX fuel allows recovery of an extra 12% of the energy in the original uranium fuel, and the recycled uranium can yield up to 22%. Pending the development of other reactor types, the spent MOX fuel is HLW and must be disposed as such. Thus, with a typical energy yield of about 34%, using MOX stretches the fuel supply by about one-third and delays its ultimate disposal by one-third of the time. MOX does not appear likely to lead to proliferation [15], although this is debated by some. The amount of fissionable $^{239}Pu_{94}$ in the spent fuel is small, and hence, the difficulty in separating it into suitable concentrations for atomic weapons is considerable. Since MOX currently yields only about 34% over and above the original energy, part of the driver towards acquiring this capability seems to be the desire to exhibit technological prowess in handling the spent fuel.

Here begin the author's problems as a writer: how to discuss the long-term effects of this waste, particularly Pu? Contending with HLW that will take 9000 years to return to background radiation is no small feat for any civilization. It has been stated [16] that, in less than 1000 years, the English language has evolved sufficiently to make Shakespeare difficult to follow, and that instructions to handle the waste must have a longer lifetime. The author can attest to the same regarding the Spanish classics, and has in a lifetime witnessed a very rapid evolution of computer languages. Hence, whatever "do not touch" instructions are issued must be capable of reaching over 9000 years. The waste is difficult to handle and to dispose of reliably, as anyone who has witnessed trees planted over buried waste and showing radioactivity can confirm.

Some other reactor types, based on Thorium (Th), are envisioned to be even less prone to proliferation than the U-based ones, and they are characterized as having less waste. Currently, they seem to be uneconomical [17,18]. Yet Th is even more abundant than U, and its most abundant isotope $^{232}Th_{90}$ can be made fertile to react completely in reactors designed to that effect, a feat to which U-based reactors cannot aspire.

RENEWABLES

What renewables lack in energy density and price they make up for in pristine environmental qualities and abundance. Other than the fossil fuel used for their construction and operation, renewables produce no CO_2. Literature searches on their environmental impact point at noise and dead birds and bats as a result of wind energy, and at toxic materials produced during the manufacture of silicon cells. However, given that the electronics industry manufactures similar semiconductors for consumers all the time, the risks associated with PV cell manufacturing appear to be small. Nothing concerning global warming or government-regulated storage seems to turn up, which are the main consequences of fossil and nuclear, respectively. Irreversibilities in the harvesting technology (friction of bearings, fluid stagnation pressure drops, electrical resistances, thermal losses) appear as heat in the environment. However, the heat would have materialized anyway had the technology been used or not.

Renewables harvesting and use neither add to nor subtract from the energy flow of our science fiction example at the beginning of the chapter. The planet gets the same flow regardless of its use. The final destination (i.e., stored in some way or dissipated into heat) does change depending on the use we make of it. The flow exists regardless; we simply would have to find a way to use it for our benefit before its quality dissipates into heat and it flows back into space.

The combustion of biomass, as given in Chapter 6, does result in large amounts of NO_x because the fuel has bound nitrogen. The cleaning procedures used for coal boilers apply, but price and complexity increase. In case we were to drift away from fossil fuels, it would be appropriate to speculate as to the rollback of CO_2 that could be expected. We can use the logistical growth models of Chapter 3 to try to project the CO_2 released to the environment when all currently known supplies of coal have been used. (This does not mean that they will be used; we simply seek to comprehend the dynamics of a very obscure phenomenon without the pretension of prediction.) The model reflecting these speculations is shown in Example 8.3.

Example 8.3 (VisSim) presents a speculation on abating CO_2.

EXAMPLE 8.3 A DYNAMIC SPECULATION ON ABATING CO_2 [VISSIM]

How much the CO_2 concentration in the atmosphere will rise before anything is done to reduce it is a matter of speculation. We assume that coal will be used as given in Chapter 3, and summarize this model as the block

LOGISTICAL GROWTH CONSUMPTION

This block calculates the yearly consumption as given in Chapter 3 and shown in Figure Example 8.3.1.

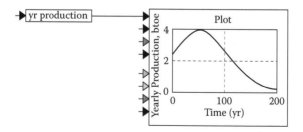

FIGURE EXAMPLE 8.3.1 Projection of coal production.

Further, from Reference 20, we have adopted a mass ratio of CO_2 to coal of 2.9, just for the purposes of this example. Also, we assume that one toe equals around 1.5 tonnes of coal. Hence, given the consumption rate, all we have to do is to multiply it by $1.5 \cdot 2.9 = 4.4$, as shown in the blocks below, to obtain the mass production rate of CO_2. It is understood that this ratio could change in the future.

$$\blacktriangleright \boxed{\text{yr production, bt}} \longrightarrow \boxed{4.4} -$$

We next assume that the "free CO_2" is the integral versus time of the difference between the CO_2 generated and that captured by plants, ocean, and perhaps responsible humans. We model the capture as a law of the type

$$CO_2 \text{removed} = k \cdot (\text{free } CO_2)^n \tag{a}$$

where the constants k and n reflect somewhat the urgency of the CO_2 problem and the ability of the species (ours, that is) to find ways to mitigate the concentration of CO_2. A large k indicates a massive sequestration effort (k today is such that the atmospheric concentration of CO_2 keeps on increasing, and whatever value it has is only due to the sequestration by oceans and plants), and n reflects the mere fact that, within limits, increased concentration results in increased absorption. Calibrating the foregoing equation is probably impossible. However, two conditions are required to determine k and n.

We focus on the past, by using the famous Mauna Loa concentration curve (Figure Example 8.3.2) [22]. The average yearly concentration increase of CO_2 is 1.25 ppm/yr in recent times. The CO_2 from coal we assume to be about 40% of this value [21], or 0.5 ppm/yr. During the Spring of the northern hemisphere, the CO_2 concentration decreases at a steep rate of about 7.5 ppm/yr. The net conversion is 0.5 ppm/yr for about 10 bt/yr of CO_2 added by coal. This yields a conversion factor of 0.05 ppm/bt.

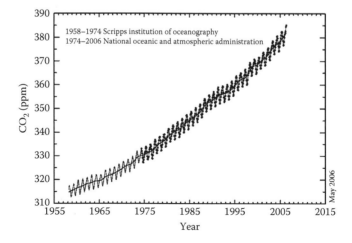

FIGURE EXAMPLE 8.3.2 CO_2 concentration in the atmosphere versus time. (From NOAA, http://celebrating200years.noaa.gov/datasers/mauna/image3b.html, 2007.)

We assume that a concentrated, international effort could eventually capture CO_2 at the current rate of release by coal, or 0.5 ppm/yr when the projected concentration would be near 400 ppm. We also assume that, if the rate of release of CO_2 were to increase by 1.5, then the capability to capture the additional increase would be developed concomitantly to reach a capture rate equivalent to 0.75 ppm/yr when the concentration raises to 500 ppm. We now have two points to calculate k and n, namely,

$$0.5 = k \cdot (400)^n$$

and

$$0.75 = k \cdot (500)^n$$

from which we obtain

$$n = -8.5 \quad \text{and} \quad k = 6.6 \cdot 10^{21}$$

These values are optimistic, in that we assume that the mere spectrum of 400 ppm would result in drastic action, and 500 ppm would certainly never be reached. However, for the species that built the pyramids, the Great Wall, rockets and a space station, windmills and nuclear power plants, this is not impossible. Hence, we input these variables in the dynamic simulation.

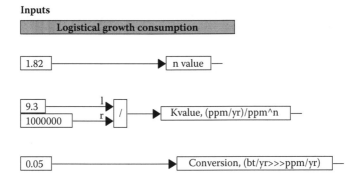

The block

$$\boxed{\text{CO2 BALANCE}}$$

simply integrates the capture equation (a) with the given values of k and n, and the initial condition of 340 ppm. The solution is displayed in Figure Example 8.3.3, along with the relevant program blocks.

The solution of the equation shows that a maximum concentration on the order of 380 ppm would be reached under the adopted scenario. If we are to depend on coal, capturing CO_2 seems like a good option.

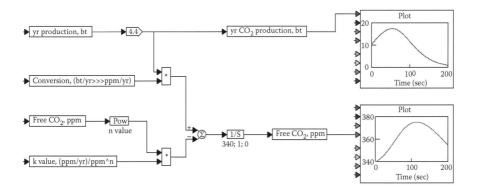

FIGURE EXAMPLE 8.3.3 Projection of CO_2 concentration if drastic action is taken.

REFERENCES

1. Feynman, R., R. Leighton, M. Sands, 1963, *The Feynman Lectures on Physics*, Vol. 1, Addison-Wesley, Reading, MA.
2. Jones J.B., R.E. Dugan 1996, *Engineering Thermodynamics*. Prentice Hall, Englewood Cliffs, NJ.
3. Cabot G., D. Vauchelles, B. Taupin, A. Boukhalfa, 2004. Experimental study of lean premixed turbulent combustion in a scale gas turbine chamber. *Experimental Thermal and Fluid Science*, Vol 28, pp. 683–690.
4. Mattison D., J.B. Jeffries, R.K. Hanson, R.R. Steeper, S. De Zilwa, J.E. Dec, M. Sjoberg, W. Hwang. 2004. In-cylinder gas temperature and water concentration measurements in HCCI engines using a multiplexed-wavelength diode-laser system: Sensor development and initial demonstration, *Applied Thermal Engineering*, Vol. 24, pp. 679–702.
5. Ferguson C.R., A. Kirpatrick. 2001. *Internal Combustion Engines*, John Wiley & Sons, New York.
6. Lave L., H. MacLean. 2002. An Environmental-economic evaluation of hybrid electric vehicles: Toyota's Prius vs. its conventional internal combustion engine corolla, *Transportation Research* Part D, Vol. 7, pp. 155–162.
7. Zhou Q., G. Huang, C. Chan. 2004. Development of an intelligent decision support system for air pollution control at coal-fired power plants. *Expert Systems with Applications*, Vol. 26, pp. 335–356.
8. Meij R., H. te Winkel. 2006. Mercury emissions from coal-fired power stations: The current state of the art in the Netherlands, *Science of the Total Environment*, Vol. 368, pp. 393–396.
9. Baba A., 2002. Assessment of radioactive contaminants in by-products from Yatagan (Mugla, Turkey) coal-fired power plant, *Environmental Geology*, Vol. 41, pp. 916–921.
10. Halloran J. 2007. Carbon-neutral economy with fossil fuel-based hydrogen energy and carbon materials, *Energy Policy*, Vol. 35, pp. 4839–4846.
11. Gutberlet H., Schallert B. 1993. Selective catalytic reduction of NO_x from coal fired power plants, *CatalysisToday*, Vol. 16, pp. 207–236.
12. Bernstein L. et al. 2007. Intergovernmental Panel on Climate Change: Fourth Assessment Report Climate Change 2007: Synthesis Report. This report can be found in http://www.ipcc.ch/ipccreports/ar4-syr.htm, or Home site for United Nations Environment Programme, http://www.ipcc.ch/index.htm.

13. Clean Air Hamilton. 2007. Fact Sheet #3, Health Effects of SO_2 and Sulphates, in http://www.cleanair.hamilton.ca/reports/factsheet3.asp.

14. World Nuclear Association. 2007. Waste Management in the Nuclear Fuel Cycle, in http://www.world-nuclear.org/info/inf04.html.

15. World Nuclear Association. 2007. Mixed Oxide Fuel (MOX), in http://www.world-nuclear.org/info/inf29.html.

16. Schobert, H. 2002. *Energy and Society*, Taylor and Francis, London.

17. World Nuclear Association. 2007. Supply of Uranium, in http://www.world-nuclear.org/info/inf75.html.

18. Uranium Information Centre Ltd. 2007. Thorium, in http://www.uic.com.au/nip67.htm.

19. Hinojosa, L. 2007. Nuclear Waste Storage and Disposal, Presentation funded by Holtec Int. to Penn State students, University Park, PA.

20. Hong, B., E. Slatick. 1994. "Carbon Dioxide Emission Factors for Coal" Energy Information Administration, *Quarterly Coal Report, January–April 1994*, DOE/EIA-0121(94/Q1. Washington, D.C.,(August 1994), pp. 1–8.

21. McAleer, M., F. Chan. 2006. Modeling trends and volatility in atmospheric carbon dioxide concentrations, *Environmental Modeling and Software*, Vol. 21, pp. 1273–1279.

22. National Oceanic and Atmospheric Administration, U.S. Department of Commerce, NOAA 200th Anniversary Web site, http://celebrating200years.noaa.gov/welcome.html, actual graph location: http://celebrating200years.noaa.gov/datasets/mauna/image3b.html.

9 Dynamic Modeling

Change is inevitable—except from a vending machine.

R.C. Gallaher in www.quotegarden.com/change.html

Although most examples in this book start from the basic equations to formulate ordinary differential equations that can be solved numerically, we offer in this chapter additional instruction on how to formulate and solve thermofluid models. For those lacking background in dynamic modeling, the material included here should be helpful.

It is clear that events happen spaced in time, and hence, dynamic modeling is a way to gauge the time scales for different phenomena. Dynamic modeling is a well-established art, with many contributions from all over the world at many different times. For an excellent treatment of the topic, the reader may consult Reference 1. We focus here on some simple aspects for modeling most thermal systems at an elementary level. Issues dealing with the nature of time, simultaneity, and entropy generation are well beyond our scope.

VARIABLES AND ELEMENTS

Dynamic modeling deals with systems and their change in time. It defines system states via variables. Systems can be mechanical, electrical, thermal, and fluid, or a combination of some or all of these. A mechanical system could be, for instance, a mass-spring-damper (Figure 9.1 (a)), described with Newton's laws. An electrical system typically includes resistances, capacities, and/or inductances (Figure 9.1 (b)), and it is described by equations reflecting voltage drops in each component and conservation of current in nodes. Thermal systems obey, in addition to the laws of mechanics and electrical circuits, the laws of thermodynamics and heat transfer and are the topic of this book, whereas fluid systems respond to pressure gradients, and their basic laws need to be used for thermal system modeling.

Each system is then composed of elements, with functions that are in many cases analogous [2]. Each element has a state identified by suitable variables. For instance, mass in mechanical systems implies inertia, for mass stores kinetic energy, and momentum flows imply changes in energy. In an electrical circuit, a capacitor (i.e., inertia) stores energy, and flow of current implies changes in the stored energy. In a thermal system, the thermal inertia is the product of the mass multiplied by the specific heat, and flow of heat implies changes in stored energy. In a flow system, a reservoir stores mass, which reflects the inertia of the system, and mass flows change

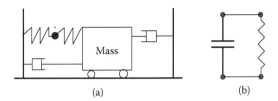

FIGURE 9.1 Examples of mechanical and electrical systems.

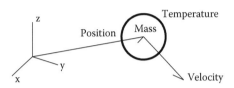

FIGURE 9.2 Flows cross the boundaries of CVs.

the storage level. Whereas elements of various systems can be identified with inertia, each element has a state that is defined by variables (Figure 9.2).

In Table 9.1a (which we will build up gradually and assign an alphabetic sequence to succeeding versions), we identify inertia and describe the element that fulfills the function in each system.

The amount of inertia itself can be the variable inside an element. To change the momentum stored in the inertia or to change the inertia itself requires a flow. For each type of system, a different flow can be identified. A flow of momentum changes the energy of a mass in a mechanical system. Current, heat flow, and mass flow change the other three inertias: electrical, thermal, and fluid, respectively. We reflect the flows in each system in Table 9.1b.

TABLE 9.1A
Element Inertia in Different Systems

Function	Mechanical	Electrical	Thermal	Fluid
Inertia	Mass (kg)	Capacitance (F)	Inertia (J/K)	Mass stored (kg)

TABLE 9.1B
Element Inertia and Flows in Different Systems

Function	Mechanical	Electrical	Thermal	Fluid
Inertia	Mass (kg)	Capacitance (F)	Inertia, (J/K)	Mass stored, (kg)
Flow	Momentum (kg*m/s)	Current (A)	Heat (J/s)	Mass flow (kg/s)

As indicated in Chapter 2, flows invariably follow favorable gradients (Figure 9.3). The fact that flows follow gradients is not only the most general source of irreversibility (reversing the flow requires reversing the gradient, which requires external, high-quality energy for reversal) but it also allows a semblance of a general formulation for all systems, in that gradients and flows can be related mathematically.

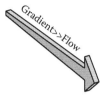

FIGURE 9.3 Flows follow gradients.

Instead of a fixed mass or inertia, we often consider control volumes (CVs), treated in Chapter 2, because CVs arise most often in the treatment of thermal systems. The elements (i.e., inertia) are still the same, but the boundaries of the CVs are crossed by flows, of mass, heat, or work (Figure 9.4).

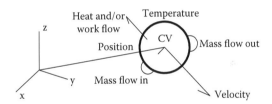

FIGURE 9.4 The boundaries of a CV are crossed by flows, which respond to gradients.

Hence, momentum flows into or out of a control volume in the direction of the applied force or via a mass flow in the direction of a pressure (i.e., force per unit area) gradient, as given by Equation 2.9 (Chapter 2):

$$\sum \vec{F} = \frac{d\vec{p}_\sigma}{dt} + \dot{m}_o \cdot \vec{V}_o - \dot{m}_i \cdot \vec{V}_i \qquad (2.9)$$

From this equation, one could visualize a force (or forces) as gradients initiating a momentum flow (Figure 9.5) as in fluid flow in a conduit responding to the force applied to a piston, or the other way around, namely, a momentum flow initiating a net force, as in a rocket or gas turbine engine (Figure 9.6).

Whereas momentum flows respond ultimately to pressure differences (point forces being somewhat of an abstraction of our minds), current I also responds to

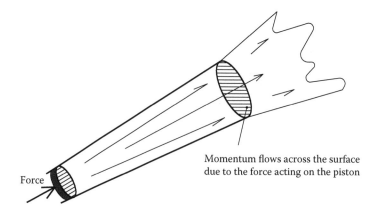

FIGURE 9.5 A force initiates a flow of momentum.

a driving force in the form of a voltage difference. When such a difference Vr is imposed across a resistor R, current I flows as given by Ohm's law (Figure 9.7).

$$I = \frac{Vr}{R} \tag{2.31}$$

We again see that a voltage gradient initiates a flow of current. Many phenomenological laws have the common denominator that flows follow gradients. In a way, they

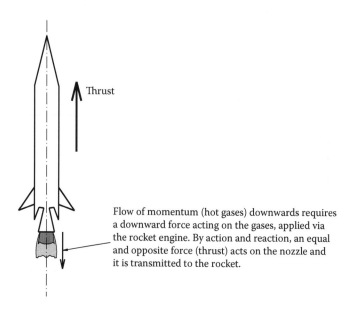

FIGURE 9.6 A momentum flow initiates forces.

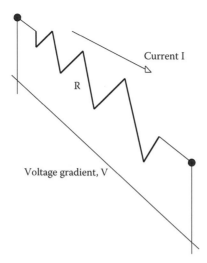

FIGURE 9.7 A voltage difference produces current.

codify the inevitable irreversibility present in all of nature. Hence, and as explained in Chapter 2, heat flows due to temperature gradients, following these laws:

$$\text{Conduction } \frac{dQ}{dt} = \frac{\Delta T}{R_{th}} = \frac{(T_1 - T_2)}{\Delta x / k \cdot A} \tag{2.37}$$

$$\text{Convection } \frac{dQ}{dt} = \frac{\Delta T}{R_{th}} = \frac{(T_1 - T_2)}{1/hc \cdot A} \tag{2.38}$$

where R stands for resistance, and it is defined differently for conduction and convection. In the case of radiation, the heat flow between two surfaces, say 1 and 2 (Figure 9.8), is proportional to the difference of the fourth power of the temperatures.

$$\text{Radiation } \frac{dQ_{12}}{dt} = \sigma \cdot \varepsilon_1 \cdot F_{A1-2} \cdot A1 \cdot \left(T_1^4 - T_2^4\right) \tag{2.27}$$

The surfaces of Figure 9.8 and Equation 2.27 must be quite special: no matter from which direction one looks at them, they emit and absorb radiation with the same intensity. In addition, radiation is an electromagnetic wave, with wavelength and frequency such as befits any self-respecting wave, even those of the ocean. Both surfaces must emit, at each wavelength, the same fraction of the maximum that could be emitted if they were perfectly black. The term σ is simply a physical (Stefan–Boltzmann) constant, ε is a property of the surface (emissivity), and F_{A1-A2} is the fraction of radiative power that reaches surface 2.

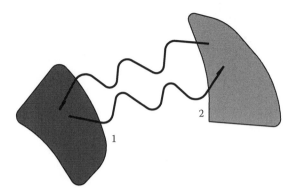

FIGURE 9.8 Two surfaces exchanging energy via radiation.

Although the form of Equation 2.27 is quite different from the ones for conduction or convection, the idea of a flow of heat responding to a temperature gradient still persists. In fact, one could think of Equation 2.27 as adopting the form

$$dQ_{12}/dt = \frac{(T_1 - T_2)}{1/\left[\sigma \cdot \varepsilon_1 \cdot F_{A1-A2} \cdot A1 \cdot (T_1 + T_2) \cdot \left(T_1^2 + T_2^2\right)\right]} = \frac{(T_1 - T_2)}{R_r}$$

where R_r stands for resistance to radiative transport and depends on the absolute temperatures.

The irreversibility built into thermal systems manifests itself also in flow systems, with flows also responding to gradients. An approximation is that flows of mass in fluid systems inside conduits respond to pressure gradients (Figure 9.9), as given by

$$\Delta p = f \cdot \frac{\rho \cdot V^2}{2} \cdot \frac{L}{Dh} \tag{2.30}$$

which can be converted to a relationship between mass flow and pressure drop using

$$\dot{m} = \rho \cdot V \cdot A$$

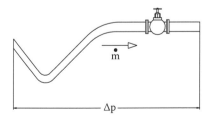

FIGURE 9.9 A pressure difference is necessary to initiate flow.

TABLE 9.1
Elements for Transient Modeling

	Mechanical	Electrical	Thermal	Fluid
Inertia	Mass (kg)	Capacitance (F)	Inertia (J/K)	Mass stored (kg)
Flow	Momentum (kg * m/s)	Current (A)	Heat (J/s)	Mass flow (kg/s)
Force	Force (N)	Voltage (V)	Temperature (K)	Pressure (Pa)

yielding then

$$\dot{m} = \left(\frac{\Delta p}{\dfrac{f}{2 \cdot \rho \cdot A^2} \cdot \dfrac{L}{D}} \right)^{1/2} = \left(\frac{\Delta p}{R_f} \right)^{1/2}$$

In the preceding equation the term R_f identifies resistance to flow. This resistance can be computed if the friction coefficient f, the length (or equivalent length) L of the path followed by the flow, the flow density ρ, the conduit area A, and diameter D are known. We now can add to our table the agent for change for each situation. Gradients of pressure, voltage, or temperature force change in each system. We label those gradients the force, and we now proceed to complete the table, namely, Table 9.1.

Since mass cannot be destroyed or created for the purposes of this text (i.e., relativistic effects are not considered), all systems follow conservation principles, and the laws derived for CVs in Chapter 2 apply to mass and energy; that is, for mass we have

$$\frac{dM(t)}{dt} = \dot{m}_i - \dot{m}_o \tag{2.7}$$

Thermal systems will undergo transient processes such that energy is conserved while being reduced in quality, which means that the entropy generation rate of the system plus environment must be greater than zero. Hence, the following equations must hold locally and for the system as well as surroundings

$$\frac{dQ}{dt} - \frac{dW_\sigma}{dt} = \frac{dE_\sigma}{dt} + \dot{m}_o \cdot \left(h + \frac{V^2}{2} + g \cdot z \right)_o - \dot{m}_i \cdot \left(h + \frac{V^2}{2} + g \cdot z \right)_i \tag{2.14}$$

$$\frac{dS_{tot}}{dt} > 0 \tag{2.17}$$

$$\text{with} \quad \frac{dS_{tot}}{dt} = \sum_\sigma \frac{ds_\sigma}{dt}$$

where, for each subsystem σ, we have

$$\frac{ds_\sigma}{dt} = +\dot{m}_i \cdot s_i - \dot{m}_o \cdot s_o + \sum_k \frac{1}{T_k} \frac{dQ_{rev}}{dt} \qquad (2.15)$$

In what follows, we will apply the cited equations to construct simple thermal system models as examples that follow the First Law of Thermodynamics for macroscopic systems, and also comply with the Second Law by adopting gradient-flow laws.

GENERAL FORMS OF COMMONLY USED LAWS

In written form, invariably we have, for the thermal systems, a relationship of the type

$$Flow = \frac{(\Delta Force)^n}{Resistance} \qquad (9.1)$$

where n can depart from 1 but it equals one in a good many cases for limited ranges of the important variables. In terms of response to flows, the general equation would be

$$Inertia \cdot \frac{dForce}{dt} = \sum Flows \qquad (9.2)$$

Note that Equations 9.1 and 9.2 are not applicable to all the systems of Table 9.1, but they work well for thermal systems.

PROGRAMMING IN VisSim

Dynamic modeling calls for building suitable ODE-based (Ordinary Differential Equations) models to ascertain the responses to various inputs. Several specialized software tools exist (VisSim©, Simulink©). We will use VisSim for these explanations. All examples so labeled are presented in the accompanying CD Rom VisSim. Some examples in Simulink can be downloaded from the book website. (http://www.crcpress.com/e_products/downloads/default.asp)

CONSTANTS AND OPERATIONS

In visual programming, functional blocks are called to reflect mathematical models. The blocks are obtained by dragging from the toolbars or from the menu "Blocks." For instance, defining constant value of 2 calls for a constant block from the menu, with the value to the block and a label as inputs in a dialog box. The dialog box is obtained by right-clicking with the mouse on the block. Reproduce this in your VisSim version:

Numerical Constant

Now, let us add another constant and display the results:

1. Define a second constant equal to 3.
2. Bring a summing junction from the toolbar.

3. Connect the constants to the summing junction.

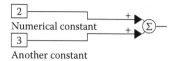

4. Add a display block.

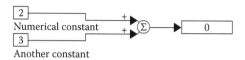

5. Simulate: click on the green triangular button, or press F5, or pull down "Simulate >> Go".

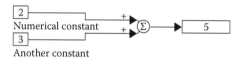

Other operations are possible: multiplication, division, and exponentiation. Try some of them on your own.

DYNAMIC SIMULATION

To model dynamically, one must invoke time. For our purposes, we use a ramp function (in the toolbar) as a clock. Dragging the other functions from the toolbar, build the following program to display the time.

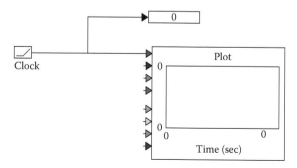

Now, when you execute the program, it displays the following:

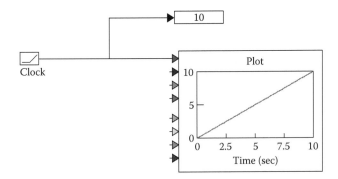

Now, change the value of the simulation time to 100 s. (Use the pull-down menu, "Simulate >> Simulation Properties," and change the end value.)

In Dynamic Simulation, there is a key to results that reflect the model. It is possible to have a model that reproduces what we intend to model, and yet the solution could be way off. This can indeed happen if the step size (pull down menu, "Simulate >> Simulation Properties > Step size") is not small enough. How can one define small enough? In using this software, the practical response is to try smaller time intervals and ascertain that the solution does not change. The curious reader might wonder as to the crucial importance of the step size for simulation.

The answer is that the step size determines the accuracy of the projection of the model into the future. The step size cannot make a "bad" model good, but, curiously, it can make a "good" model bad. Here is (roughly) how it works. Computers solve ODEs numerically. Hence, one starts with a differential equation:

$$\frac{dy}{dt} = f(t) \tag{9.3}$$

and approximates the derivative as a ratio of finite differences, namely,

$$\frac{\Delta y}{\Delta t} = f(t)$$

from which one obtains

$$\Delta y = f(t) \cdot \Delta t \tag{9.4}$$

Now, if one knows the value of y_n at some initial time t_n, one can approximate the value at an instant t_{n+1} as indicated by Equation 9.4 and Figure 9.10:

$$\Delta y = y_{n+1} - y_n = f(t_n) \cdot \Delta t \tag{9.5}$$

and casting the future value of y from the foregoing equation, we get

$$y_{n+1} = y_n + f(t_n) \cdot \Delta t \tag{9.6}$$

It is clear from Figure 9.10 that, as the interval Δt becomes smaller, the error (i.e., the difference between the true value and the numerical estimate $[F(t) - y_{n+1}]$) decreases. Decreasing Δt is not the panacea that it promises to be because increasing the

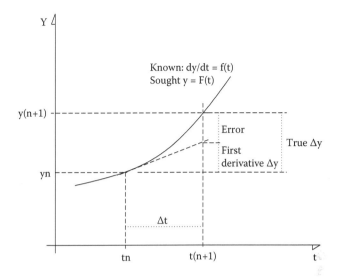

FIGURE 9.10 The step size determines the quality of the numerical approximation.

number of steps increases the computing time as more equations like Equation 9.5 need to be solved to reach a certain time. In addition, small Δts, even for the patient, have a limit: the smallest number that the computers can recognize, or rather, the difference between the smallest recognizable number and zero. This difference depends on the type of computer, and it is usually called the round-off error.

Because of the error in approximation (Figure 9.10) and the round-off error, other methods that reduce the approximation error and are more stable when using small Δts have been devised. VisSim offers Runge Kutta methods, which incorporate more points in the evaluation than t_n and y_n and are in general acceptable practice for thermal system integration. Investigate the various integration methods offered by the software under the "Simulate >> Simulation Properties >> Integration methods" in the pull-down menu.

In what follows, we develop a few examples so as to help the beginner develop programming skills. After completing these examples, the reader will be able to program any ordinary differential equation representing thermal systems.

Example 9.1 (VisSim) is included in the CD ROM.

EXAMPLE 9.1 THE RATE OF ACCUMULATION OF MASS [VISSIM]

(*Note:* RC stands for Right Click of the mouse; LC stands for Left Click.)

A lake receives a flow $\dot{m}_i$ from all the tributaries. The hydroelectric dam flow for power production is $\dot{m}_o$, and there is a leakage rate to the ground of $\dot{m}_{loss}$. Determine whether the lake level is increasing or decreasing, and at what rate.

$$\dot{m}_i = 5.6 \cdot 10^4 \text{ kg/s} \quad \dot{m}_o = 3.2 \cdot 10^4 \text{ kg/s} \quad \dot{m}_{loss} = 2.8 \cdot 10^3 \text{ kg/s}$$

Also give the time needed to take the lake inventory at the starting time from M_o to $(1.2 \cdot M_o)$.

Solution

The equation for mass conservation, Equation 2.7, reads:

$$\frac{dM(t)}{dt} = \dot{m}_i - (\dot{m}_o + \dot{m}_{loss}) \tag{a}$$

which naturally means that the rate of mass accumulation is the difference between the inlet and exit flows. In VisSim, we define the three flows.

1. Drag from the toolbar a constant block, $\boxed{1}$—
2. RC on the block. A dialog window appears. Change the value from 1 to, say, 56,000, the value of the input flow. Close the window.
3. From toolbar (or from the pull-down menu Blocks >> annotation >> variable) drag a variable (identified as "var" in red) block. Place close and to the right of the constant, 56,000 block.
4. Join the output of the constant block to the input of the variable block. Change the name of the variable block by RCing on the block and entering the new name in the window. Be sure to include units so as to remain consistent in your programming.
5. Define the output and leakage blocks. Also, calling a "label" block, and right-clicking on it, define a region for inputs.

At this stage, your program should appear as in Figure Example 9.1.1.
 We now implement the ODE (a)

1. From the toolbar, we drag a summing junction.
2. We drag the symbol $\xrightarrow{+}$ from the toolbar and click on the summing junction. A new input appears.

3. We now copy from the input list the variables for the given flows, and after pasting each, we connect them as inputs. Since the power plant and leak flows are negative, we change the sign of the corresponding summing junction by holding the control key and RCing on the mouse. Figure Example 9.1.2 shows the state of your program.
4. The output from the summing junction is the rate dM/dt of Equation (a). We define a variable dM/dt by connecting the output to it.
5. To the output of dM/dt we now connect a visual display from the toolbar to show Figure Example 9.1.3.

Inputs Flows

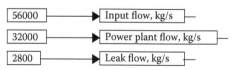

FIGURE EXAMPLE 9.1.1 Data input.

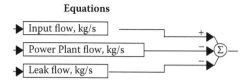

FIGURE EXAMPLE 9.1.2 Summing junction with variables connected.

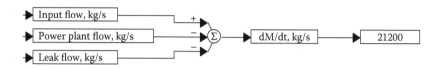

FIGURE EXAMPLE 9.1.3 Display the value of *dM/dt*.

Integrating Equation (a),

1. From the Blocks pull-down menu, we click on Integration >> integrator.
   → $\boxed{1/S}$
     $IC{:}0; ID{:}0$
2. RCing on the integrator opens a dialog window, and the initial condition (M_o) can now be entered.
3. The output of *dM/dt* is connected to the integrator input. The program now looks as in Figure Example 9.1.4.

Integration and results

1. We now prepare the simulation by using the pull-down menu Simulate >>> Simulation properties. We set the end time to 1000, and keep the time step to 0.1.
2. The integrator output is connected to a plot (you can drag one from the toolbar) (blue trace).
3. A signal of value 1.2 of M_o is generated using toolbar blocks and fed into the plot (hollow dot trace).

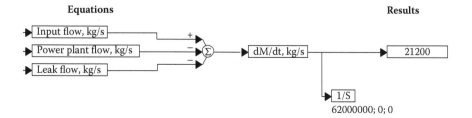

FIGURE EXAMPLE 9.1.4 Integrating the value of *dM/dt* with the given initial condition.

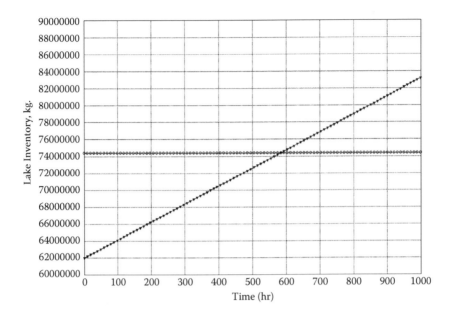

FIGURE EXAMPLE 9.1.5 Integrating the value of dM/dt with the given initial condition.

4. The dot and hollow dot traces intersect, and at that time (590 hr), the desired capacity is reached. By pressing RC after enlarging the plot by LCing on the top right square, a dialog box appears to format the plot, and one can trace the lines to the intersection point. The full-screen plot is shown in Figure Example 9.1.5.

Example 9.2 (VisSim) discusses rocket fuel mass.

EXAMPLE 9.2 ROCKET FUEL MASS [VISSIM]

(*Note:* RC stands for Right Click of the mouse; LC stands for Left Click.)

A rocket (Figure Example 9.2.1) fires its engine, expelling mass according to the following schedule:

From $to = 0$ s to $t1 = 5.56$ s, $\dot{m}_{o1} = 0.1 \cdot t + 0.01 \cdot e^{0.1t}$

For $t > 5.56$ s, $\dot{m}_{o2} = e^{-0.1t}$

Calculate the total mass ejected in 50 s.

FIGURE EXAMPLE 9.2.1 Rocket schematic.

Solution

The equation to model is

$$\frac{dM(t)}{dt} = \dot{m}_{o1} \tag{a}$$

Each of the discharge functions $\dot{m}_{o1}$ and $\dot{m}_{o2}$ is easy to build, and they are shown in Figure Example 9.2.2. (In the VisSim Example 9.2, RC on each block to see how the equations were constructed.)

To obtain the called-for function at each point (i.e., one for $\dot{m}_{o1}$ and one for $\dot{m}_{o2}$) in time, we multiply each by a factor that is equal to one only when the function is active. This can be done using two Boolean blocks (Figure Example 9.2.3) for selection of one input. (To see the blocks, RC on the Select block.)

When the clock time is less than 5.56 s, the upper block is one and the lower block is zero. When the time exceeds 5.56 s, the upper block switches to zero and the lower to one. Hence, each of these functions is multiplied by the output of the corresponding Boolean block. Each function is now transformed into a compound block. Select

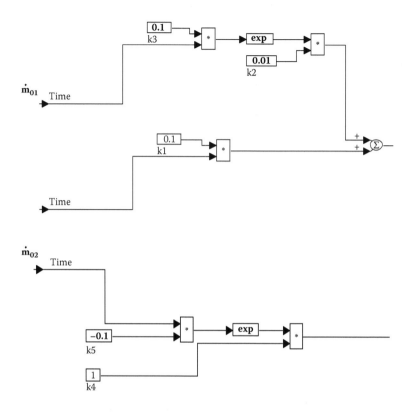

FIGURE EXAMPLE 9.2.2 Equations for $\dot{m}_{o1}$ and $\dot{m}_{o2}$.

the blocks that will form the compound block. Pull down the Edit menu and choose "Create Compound Block." A dialog window appears, and you can give the block any desired name by typing it now. After typing the name, click OK, and the block is formed. To access the block, place the cursor on it and LC the mouse.

We have created by now three compound blocks: one for $\dot{m}_{o1}$, one for $\dot{m}_{o2}$, and one for the Boolean set of blocks ("Select" Figures 9.2.3 and 9.2.4) that activates either $\dot{m}_{o1}$ or $\dot{m}_{o2}$. After each of the three block diagrams is made into a compound block (this is a block that contains all the individual blocks of the function), we have then the following program (Figure Example 9.2.4).

As shown in Figure Example 9.2.4, the integral of the mass flow rate over 50 s indicates that 7.2 kg have been consumed.

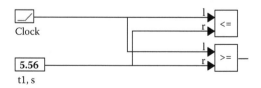

FIGURE EXAMPLE 9.2.3 Blocks for selection of one input.

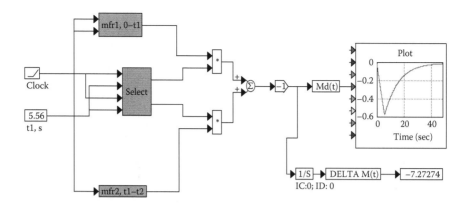

FIGURE EXAMPLE 9.2.4 Completed program. ($\dot{m}_{o1}$ and $\dot{m}_{o2}$ are identified as mfr1 and mfr2 in the program.)

Example 9.3 (VisSim) discusses heat storage.

EXAMPLE 9.3 MASSIVE STORAGE WALL [VISSIM]

Massive walls are used to store thermal energy. No innovation here, massive construction and light-color paint have been common practice in warm areas of the world for centuries in order to keep the ambient cool during a sunny day. We now study a more recent application—that of a wall that is heated during the day by solar radiation and is cooled at night in order to keep the ambient at a reasonably high level.

During the day, the wall is heated by air from a collector to Tinit. We define Tinit in the inputs, along with the wall mass and minimum acceptable wall temperature after heat has been removed at a rate of 7 kW (Figure Example 9.3.1).

The crucial equation is that of the heat balance,

$$\frac{dQ}{dt} - \frac{dW_\sigma}{dt} = \frac{dE_\sigma}{dt} + \dot{m}_o \cdot \left(h + \frac{V^2}{2} + g \cdot z \right)_o - \dot{m}_i \cdot \left(h + \frac{V^2}{2} + g \cdot z \right)_i \quad (2.14)$$

which readily simplifies, in the absence of work and for the given heat loss, to

$$\frac{dQ}{dt} = \frac{dE_\sigma}{dt} \quad (a)$$

Inputs

FIGURE EXAMPLE 9.3.1 Inputs for heat storage wall.

Equations

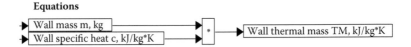

FIGURE EXAMPLE 9.3.2 Calculation of TM.

Since the rate of change of energy stored is the thermal inertia multiplied by the rate of change of temperature, we have, for the thermal inertia *TM*,

$$TM = c \cdot m \tag{b}$$

where c is the specific heat and m is the wall mass. We calculate *TM* (Equation (b)) (Figure Example 9.3.2) and also implement the solution to the ODE (Equation (a)) in the form:

$$\frac{dQ}{dt} = TM \cdot \frac{dT_{wall}}{dt}$$

In analog programming, the equation with the given initial condition is shown in Figure Example 9.3.3, which also shows the temporal variation of wall temperature. In 30,000 s (about 8 hr), the temperature decreases from 363 to 333 K, and the wall delivers a total of 90,000 kJ. Although integration is really not necessary in this straightforward example, it was included in order to better illustrate the procedures.

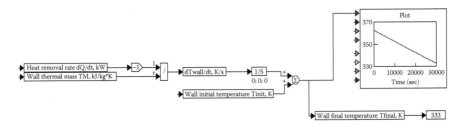

FIGURE EXAMPLE 9.3.3 Integration of Equation (b), showing the temperature decrease versus time.

REFERENCES

1. Kulakowski, B. T., J. Gardner, J. Shearer, 2007. *Dynamic Modeling and Control of Engineering Systems*, Cambridge University Press, New York.
2. Lewis, J., 1994. *Modeling Engineering Systems*, High Text Publications, California.

10 The Future
A Moving Green Target

Living is a constant process of deciding what we are going to do.

—**Ortega y Gasset**

We have focused so far on existing technology even if some of it has not penetrated the market to a large extent. Much research is being done in many areas of energy conversion and end use. Although a summary hardly does justice to all this effort, this chapter is but a summary of what might make a difference in the future. We rely heavily, as in Chapter 3, on the opinions of many to ascertain what the energy future might entail.

THE FUTURE AND CRITERIA

Writing about the future with any certainty is impossible. Energy is inextricably linked to discovery, production, distribution, use, finance, and government policy of many nations. Each energy source that gains acceptance develops a sector (or sectors) that comprises developers, users and germane technology. To predict how all these sectors will interact is unattainable. What is clear is that the more accumulated knowledge and experience each sector has, the more intensely will the sector persist in advocating a future within which its role is undisputable or crucial. The cases that do not fit this observation are rare. Technologies become obsolete and get replaced with versions more suited to the circumstances and energy availability of each generation, but the sector, if strong enough, persists. For instance, coal has survived deplorable direct-heating applications, primitive boilers, reciprocating engines for sea and ground transportation, even coal liquefaction. Perhaps in the future, fuels from coal will activate aircraft. This author cannot find any sector (or within a sector a technology) that spontaneously dissolved to let a competing sector or technology fill its place. It might have happened, but if it did, it probably was due more to chance than to a coincidence of purpose.

So, what can this chapter offer the reader? We will endeavor to project the possibilities of technologies for different sectors based on a set of broad criteria. The chosen criteria attempt to factor in the time variable in our decisions, but they do not amount to life-cycle analysis, whether conventional or otherwise. The reader will not find here any economic projections; only qualitative ones. In particular, we endeavor to factor the three criteria that, for the foreseeable future, seem permanent: the First Law, the Second Law, and legacy issues.

The First Law criteria concern abundance of the energy resource. Because local variations are evident, we factor them into the discussion when possible, but generally we stay within the global framework of Chapter 3. The Second Law criteria concern efficiency. As noted throughout this work but especially in Chapter 7, all processes (cyclic or not, chemically reacting or not) dissipate some mechanical energy into heat. It is difficult to quantify in advance how much will be dissipated. Projections of dissipation rates are typically the product of studies of variable empirical content, a content which is never absent. What matters is our contention that, if the dissipative losses are too large, the First Law constraints will reduce the penetration of the technology. Legacy criteria are those that attempt to keep future generations from paying for the poor judgment of the previous ones when the latter misjudged the irreversibility or time scale associated with an energy process.

COAL: MANY POSSIBLE FUTURES

From the viewpoint of abundance as studied in Chapter 3, the coal supply (world and U.S.) is ample. However, it is not a flexible fuel in terms of its convertibility to gas or liquid fuels. The gasification of coal has a long history, and whenever other fuels seem to (or actually do) become scarce, coal gasification offers a solution. It is much easier to transport and combust clean gases than coal, and the gasification allows, in theory, removal of all contaminants *in situ*.

COAL GASIFICATION

This technology invariably calls for a reactor under pressure and at elevated temperature in which coal particles react with air or oxygen. The principle [1] relies on using exothermic reactions to sustain the formation of desirable compounds that might be endothermic. It is a particularity of this technology to resort to a cheap source of hydrogen, namely, water. Hence, large amounts of water are required for gasification. The reaction of carbon and oxygen

$$C + O_2 \rightarrow CO_2 \qquad \Delta H = 393790 \ \text{kJ/(kg} \cdot \text{mol)}$$

is exothermic, and it can in principle drive the endothermic reaction

$$C + CO_2 \rightarrow 2CO \qquad \Delta H = -172580 \ \text{kJ/(kg} \cdot \text{mol)}$$

The addition of steam to the mixture enables another two reactions:

$$C + H_2O \rightarrow CO + H_2 \qquad \text{(Synthesis Gas, endothermic)}$$

$$C + 2H_2 \rightarrow CH_4 \qquad \text{(Hydrogenation, exothermic)}$$

When the system is conditioned to allow all these reactions to proceed simultaneously, then the gas obtained will contain CO, H_2, and CH_4 in varying degrees, as well as CO_2, N_2, and other compounds. The gases can also react, resulting in a composition that reflects the reactor conditions and kinetics. Thus, CO may react with either H_2O or H_2 to result in more H_2 or methane, respectively. The gases thus obtained are combustible, and may be used for multiple purposes in technologies similar to those described in this text for gaseous fuels. The broad rules of chemical equilibrium establish that, at high temperatures, the formation of CO will be favored, whereas high pressures foster the formation of CO_2 and CH_4.

In trying to establish the perspectives of coal gasification, it is fair to say that, unless CO_2 and other products (such as Hg and particulates) could be removed at an acceptable energetic (and financial) cost, then its usefulness might be rather limited. A recent publication [2] projects prices for electricity derived from IGCC with no carbon capture, with 80% carbon capture, and for electricity from a natural-gas-activated combined cycle with no carbon capture (NGCC). The latter cycle was conceptually described in Chapter 4. For IGCC, the projected power prices range from 5.09 to 6.67 c/kW · hr, the upper limit corresponding to 80% carbon capture. For comparison, the NGCC electricity cost is, according to the same authors, 3.54 c/kW · hr, with greater carbon emissions but larger thermal efficiencies than with IGCC. From this perspective, the choices are clear: if climate change is indeed a reality and not the mirage that the collective actions of most societies outside Europe would deem it to be, then IGCC with carbon capture is an urgent necessity. However, at the outlined price for electricity production, other options not as benign to the environment might prevail.

LIQUID FUELS FROM COAL

The economic urgency may very well favor coal to liquids in the future, as it did in war situations in the past. A recent report [3] communicates the fact that a large corporation has produced J-A1 fuel from coal via a proprietary process, and that such fuel has been approved for use by airlines. The energy density of liquid fuels is currently indispensable for flight, and it remains to be seen whether it can be dispensed with for individual long-range transportation. Coal liquefaction implies the formation of bonds present in liquid molecules, namely, carbon–hydrogen (C–H) bonds, and of carbon–carbon (C–C) bonds. The longer the carbon chains are, the lower the vapor pressure of the liquid will be, at a given temperature. Bituminous coals [2] have a H/C ratio of 0.8, whereas the ratio for gasoline is 1.9. Hence, liquefaction implies increasing the H content of coal. In direct coal liquefaction, coal reacts with H at high pressures and temperatures, with a solvent acting as H donor and a catalyst being present in some cases. A form of this technology is slated for implementation in China. In indirect coal gasification, the syngas from coal gasification (particles having been removed), is cooled and fed to a synthesis reactor to produce liquid hydrocarbons similar to those distilled from petroleum. This is the Fischer–Tropsch process, used in South Africa to produce gasoline from coal. According to Reference 4, direct liquefaction processes are projected to capture 55% of the energy in the coal versus 45% for Fischer–Tropsch.

The key issue in coal liquefaction (and to a lesser extent for syngas production) is that of hydrogen: where will it come from? Water is the only practical source of hydrogen, and water, especially potable water or reasonably clean water, becomes a more and more expensive commodity as population and development increase. Because the hydrogen returns to water upon combustion, the issue is not one of scarcity but of water availability in conjunction with the coal supply. In this sense, some areas of the world are better endowed than others, and hence, better suited to coal hydrogenation.

CO_2 CAPTURE AND ITS POSSIBILITIES

Biomass captures CO_2 through photosynthesis with solar energy input. No other process captures solar energy, produces vital O_2 along the way, and stores CO_2 in useful, reduced form. The idea of controlling CO_2 via land and crop management has been addressed, and it seems that real possibilities (beyond what occurs naturally) exist [5,6]. However, outside of CO_2 control via biomass growth, what is the real possibility of CO_2 capture?

A recent publication [7] suggests that capture of CO_2 could become standard practice if leading economies were to adopt pre- or postcombustion capture, either in new power plants or retrofitting existing ones. Developing nations would then adopt the technologies after the learning cycle is complete. The suggested time frame would be around 2020 for the process of adoption to be in full fledge. Desirable as these developments would be, they require a driver that this author does not see uniformly distributed in populations and governments. Besides, the capture of CO_2, as noted with reference to IGCC [2], entails additional energy and disposal costs. Hence, it may be difficult to impose on all economies, which makes it harder to impose on the first one.

Most of the CO_2 in the world [8] is produced by the power/heat industry and by transportation. Based on the data of Reference 8, it is clear that the United States and China produced in 2004 about 10 Gt of CO_2 versus the world total of 26.6 Gt. Emissions per capita vary widely: they were for China, in 2004, one-fifth of that of North America. Controlling emissions in regions of high production would go a long way toward reducing CO_2 in the atmosphere. Other countries contribute much less CO_2: Japan, with the second GDP of the world, produces one-tenth of the CO_2 of the United States and China combined, while relying on nuclear for 24% of its electric power production. Clearly, energy-efficient technology is important in Japan since it enables low emissions and large economic output. Renewables, as highlighted by the case of Brazil in the form of biofuels and hydro [8], can help to realize a productive economy with reduced CO_2 discharge.

Because Japan and Brazil produce less CO_2 for their economic output than other nations, they are more buffered from fossil fuel price increases than economies that depend less on energy efficiency and renewables. Those two countries could serve as the models sought by Reference 7. Perhaps CO_2 emissions will be reduced as price constraints force the major contributors to adopt more efficient technologies that reduce consumption. Decreasing demand could result in lower prices that would again fuel development of CO_2-emitting technology, but we count on a certain maturity of the international community not to give up technological gains that result in more efficient economies.

OIL AND GAS

ADVANCED POWERTRAINS

As the author writes these thoughts, the price of gasoline in the United States continues to increase, and the projections of Chapter 3 (done in 2006) suddenly become significant. If indeed we were in a peak then, and if indeed the peak oil theory is more than a fad, then the prices we see are but the result of declining production. Because production has not decreased but its rate of increase has become smaller [9], it is hard to ascertain whether we are in a peak transition or not. What is clear is that oil reserves and prices remain high, although reserves decreased only slightly in 2006 [10].

Oil provides jet fuel for air travel. Aircraft will cope with higher fuel prices and increased demand by developing better and better turbofans. Turbofans are jet engines that generate thrust by efficiently moving large amounts of air with optimized ducted fans. The prime mover activating the fan is a gas turbine. Because no other prime mover has weight-to-power ratios comparable to those of turbofans, their primacy in air travel is far from over. Also, the energy densities of liquid hydrocarbons, when coupled with turbofans, make a powerful association that no current technology in the research stage is projected to beat. Certainly, liquefied coal could power air travel, but at reduced fuel production efficiencies.

The internal combustion engine (ICE), described in Chapter 4 as one of the most commonly used prime movers, uses gasoline. ICEs have been one component of the strongest fuel-technology-user triad of the 21st century. Gasoline, ICEs, and individual transportation have defined the lives and livelihood of multitudes. Naturally, as oil prices change, the key question is how the technology will evolve, or whether it will disappear altogether, no longer able to provide individual transportation affordably enough as to sustain an industry. Competing technologies to ICEs are hybrid powertrains (i.e., powertrains with an ICE and electric motor, with batteries for storage) and electric powertrains (i.e., batteries and electric motors). Hybrids imply continued existence of ICE engines. Because hybrid powertrains can have recuperative braking, their energy efficiency could be, in principle, superior to that of ICEs. There is increased complexity and price associated with hybrids, which may, in fact, give rise to the adoption of yet another powertrain, the diesel engine. The latter attains mileage comparable to that of most hybrid systems.

The electric powertrain relies on batteries and on an electrical motor. The batteries obtain their charge from a wall outlet. In yet another version, a small on-board gasoline engine can also serve the batteries. Let us call this latter version a plug-in hybrid. The key difference from current ICEs is that the engine simply charges batteries, as opposed to driving the car directly. Will such powertrains deliver the same utility with less energy? In more general terms, if a car runs on power from a wall outlet (or from, say, diesel fuel for charging), will some of the clearly identified problems regarding energy supplies and CO_2 be alleviated?

The answer is complex because it largely depends on how well the technology is implemented and used. Yet, the discussions of Chapter 7 regarding the First Law warrant broad projections. Since electrical production from natural gas has a (current) maximum efficiency of 0.6 and a (current) national average of around 0.3, then the average conversion from fuel to power on the wheels cannot exceed 0.3. Incorporating the battery efficiency and electric motor efficiency (assume 0.8 each) would result in a combined efficiency, fuel to wheels, of 0.2. Recuperative braking could enhance this efficiency, the extent of improvement depending on terrain and traffic conditions. If one assumes an improvement from 0.2 to 0.25 due to recuperative braking, then the chain efficiency is comparable to that of existing

ICEs. The only crucial difference is one of fuel substitution, for cars would then run on a mixture of coal, renewables, and nuclear. Whether the grid use would be rational (i.e., most charging occurring at night) is hard to say, but straining the grid or production facilities with demand from hybrid cars during peak hours is a clear possibility that probably would sort itself out via pricing.

In summary, the future of individual transportation is linked to diesel and plug-in hybrids in varying degrees. Of course, an invention that increased the compression ratio of ICEs could go a long way toward modifying this projection.

COGENERATION CYCLES

The world has an ample supply of natural gas (not as plentiful in the United States), the only question being how efficiently it will be used in the future. Natural gas is a versatile fuel that can be used in technologies of high efficiency. Gas turbines, in cogeneration cycles, will likely continue to improve, resulting in performance gains. Higher inlet temperatures and pressure ratios for gas turbines are just a matter of time, given the strong drivers behind R&D for this technology. ICEs, already powered by natural gas in many areas, could increase in number, but the efficiency of use is not as high as in cogeneration cycles, scaling up is difficult, and hence, their expansion is unlikely if fuel prices remain high. Natural gas use in the residential and commercial sectors should use higher-efficiency technology. For instance, fuel cells of increased efficiency and reduced prices could gain acceptance. However, power production performance should approach that of gas turbine combined cycles before widespread diffusion can be projected.

MOVING FUELS AROUND

Some structures, old or new, encompass such engineering knowledge and/or have a scale that denotes such inspiration and labor that they are of true historical significance. One such system is the one built to transport energy all over the world. As of 2006, 64% of the oil produced [11] and 26% of the natural gas produced [12] was subsequently traded internationally. A large and complex infrastructure consisting of pipelines, tankers, and ancillary installations and equipment has been created to sustain this trade.

Some 3500 oil tankers, about 200 liquefied natural gas (LNG) ships, and numerous international and domestic pipelines comprise a conveyance network of staggering scope. Accidents are rare, but they happen with noticeable environmental impact and also occasionally with loss of life. This trend of moving fuels responding to a clear economic driver will not abate but rather will keep on increasing, especially for natural gas. The investment in knowledge and hardware is too large for this structure to come to a spontaneous halt. Shifting production to renewable power might abate some of the international trade, hopefully accompanied by more price stability. In the absence of a clear intellectual driver for renewables, the profitable fossil energy trading network will likely expand in the future.

NUCLEAR ENERGY

In projecting what nuclear might bring to the energy future, it is necessary to consider power plants and waste disposal. There is currently a trend to inform the public about the undeniable possibilities of nuclear energy [13]. From Reference 13, one could conclude that emotions seem to cloud issues. It is the issues, and only those issues which are treated objectively, that we attempt to summarize in what follows. Nuclear power plants are either of the pressurized water reactor (PWR) type (briefly described in Chapter 4), or of the boiling water reactor (BWR) type, which differs from the PWR design in that the water is evaporated in the reactor and flows directly through the turbine, that is, the intermediate evaporator (boiler) of Figure 4.18 (Chapter 4) is eliminated. Both designs are used throughout the world; 439 power plants have an average capacity of some 820 MWe [14]. Except for a few accidents, power generation has proceeded safely and reliably. The United States has 104 operating nuclear reactors.

NEW REACTORS

New designs have been developed with a variety of funding sources, invariably with governmental support in one measure or another. The central idea of these designs is the use of passive safety techniques, such as procedures that, in case of malfunction, correct themselves automatically, relying on a naturally occurring phenomenon such as natural convection or decreasing reactivity with increasing temperatures. Equipment manufacturers have a number of new designs in varying stages of implementation. These designs promise enhanced safety and efficiency [15]. Costs of new designs are projected to be in the range 1500–2000 $/kW·hr, with thermal efficiencies in the 0.35 to 0.38 range. Two new designs for light-water reactors are the advanced boiling water reactor (ABWR) and the AP-1000; both of them have passive safety features that met approval of regulatory agencies. There are also new designs for heavy-water reactors that have enhanced safety features. Of particular interest are those designs based on thorium (Th) [16], which is converted in the reactor to the fissile isotope U^{233} by slow neutrons. Thorium reactors, hence, do not necessarily breed Pu^{239} (of atomic weaponry fame), but short-lived Pu^{238}, which reduces the danger of proliferation [16]. The element Th is vastly more abundant (3 to 4 times) than U in the Earth's crust, and the United States holds the second largest reserves. All the Th atoms are fertile, as opposed to a small percentage in the U ore.

Another advanced technology is that of gas-cooled reactors, which utilize fuel pellets with a moderator coating that prevents high temperature excursions by absorbing neutrons more efficiently as it heats up. Hence, the reactor is passively safe. The coolant agent is He, and the gas is expanded in a gas turbine. Because the reactor operates under the pressures of a gas (He), which, unlike liquids, can expand in case of vessel rupture, safety is increased by building the reactor underground.

Regarding power plants, then, the future of nuclear energy is ample. With governmental support, the new designs will ease the pressure on fossil fuels to meet the energy demands of a growing population with growing expectations. This is particularly the case for cycles based on Th. This transition to nuclear energy is ongoing, and will continue as development searches for energy.

WASTE DISPOSAL

The other issue surrounding nuclear energy is waste disposal. Nuclear power plants embody so much specialized knowledge that, in a way, nuclear energy is comparable to space or submarine technology: only comparatively few professionals can work on the design and construction of nuclear power plants. A parallel can be drawn between the waste generated by combustion technology and by nuclear technology, even though the degree of specialization differs. Initial development of both technologies ignored the effects that by-products have in numerous realms, although the time scales are vastly different for fossil than for nuclear fuels.

Wastes can be buried in various forms, in stable geological formations and inside canisters of several specialized materials. An alternative to burial (which implies transport if the burial is to occur at only one site), consists of waste destruction/utilization in a Fast Neutron Reactor, a technology still not fully developed. The only other alternative is to store nuclear waste in nuclear power plants for many hundreds of years. (High-level wastes return to background radiation levels in 9000 years, although the actinides do so in about 600 years [17].) This solution raises the possibility of large amounts of waste distributed over wide areas. Management of this situation requires vast resources, for either gathering the wastes or keeping them dispersed requires coordination, safeguarding, and complicated logistics for long time spans. In the ultimate analysis, it is a legacy question, similar in principle to that of CO_2 in the atmosphere; it is a highly irreversible situation, and those who enjoyed the energy will eventually leave the sequels for someone else to deal with.

Nuclear waste is dangerous for a variety of reasons, and hence, no one really wants it (or should want it), although the electric power is clearly sought by many nations. Therefore, in one way or the other, governments must monitor and control the fate of the wastes in order to avoid danger to the inhabitants of the planet. Unless the waste problem is satisfactorily solved at the international level, nuclear energy may never live up to its full potential, and may pose a myriad of problems to future generations.

RENEWABLES

In terms of future, renewables have an ample one, and the only question is whether such will always be the case, or alternatively, we will find our way to exploiting their full potential. Renewable energies have what other forms of energy lack, that is, when they are captured and used, no additional energy is released. Simply put, out of an incoming stream of energy, the users derive what they need. Hence, no thermal or other forms of pollution arise, other than that incurred when manufacturing the harvesting technology. The latter affects all forms of energy, renewable or not, since the mining and extraction of fuels as well as the production facilities need implementation that demands energy.

NEW SOLAR AND WIND CONCEPTS

The most rational conversion of solar energy is into electricity. However, we saw (Chapter 5) that efficiencies of photovoltaic technology are low, even if sufficient to power average households in areas of the Midwest. The efficiencies of conversion into thermal are high, but thermal energy needs to be converted to electric power via a thermal cycle of efficiency limited by Carnot. Solar implies storage, but only for those seeking complete independence from the grid or from a fuel distribution network.

Whereas the cost of photovoltaic and other solar cells will come down as economics of scale and research uncover new approaches, such as dye-sensitized and plastic solar cells, the storage issue is much more uncertain. Batteries are expensive, they impose a burden on efficiency, and they increase the investment of material and energy necessary to implement the technology. They also increase the footprint of any system. Hence, generalized energy storage coupled with renewable may be hard to achieve.

Perhaps, as the harvesting of solar power (responding to necessity and acceptable prices) becomes gradually widespread, the existing fossil and nuclear power generation and distribution facilities will fill the role of baseline and night power. Eventually, as both solar and fossil/nuclear technologies evolve, conventional fuels could fill out the winter/night transients via generation of fast response (such as that possible with gas turbines [18,19]), while solar power could meet peak demand during summer periods with increased reliability and reduced costs.

The likely advent of plug-in hybrids (already discussed) could mean the addition of storage for renewables. The combined storage capacity of a large number of cars could indeed furnish renewables with a reliable storage vector that is currently hard for most homeowners to acquire. Perhaps automobile factories will incorporate means to discharge the batteries into the grid when conditions warrant it, reducing the electric bill of the user.

Wind energy seems to be on the track of ever-increasing wind turbine capacity. Better controls are allowing windfarms to increase the efficiency of energy harvesting by taking advantage of wind transients and improved aerodynamic blades. No new technologies based on different principles seem to be under development for wind, but perhaps the maintenance needs of a distributed energy network, such as the one called for by wind, will call for radically new concepts without large moving parts.

A CASE FOR BIOFUELS: NOT ALWAYS OR EVERYWHERE

Liquid fuels, more in demand than ever, can be made from biomass. However, the consequences of such manufacture are hotly debated. Fuel production uses surface area that could be used to produce food, a consideration of the utmost importance. A recent article [20] serves as an example of the conundrum: corn ethanol hardly produces any positive energy return—a ratio of investment to yield of 1.3—whereas switchgrass promises a yield from 3 to 7. As the use of corn ethanol increases, so does the price of corn. These are sobering facts, but this chapter (and this text, really) is aimed at discussing what could really improve the thorny energy/pollution situation that we have worked ourselves into.

As explained in Chapter 5, biomass captures only 1/15 of the energy that a solar cell is capable of capturing, and then the biomass energy must still undergo a number of transformations before it can be used for transportation. Such generalized use, should it come true, is projected to demand about half of the current agricultural land [21]. Reference 21 questions the carbon removal potential of the transition to widespread biofuel use. Whereas these projections have large uncertainties, what is clear is the low (if indispensable for oxygen production) efficiency of biomass for energy purposes. Regardless of the technology adopted, the bottleneck is the efficiency of photosynthesis.

We noted earlier the case of Brazil that is capable of energy independence based on sugarcane ethanol. Large energy yields have been reported for the way this conversion is made (from 5 to 8) [22]. In any case, one of the distinguishing features of Brazil is the ample availability of land with good solar irradiation. Perhaps the best use of biomass occurs when there is a waste product or by-product from which a fuel can be derived. The combustion of biomass is challenging, but a number of organizations worldwide serving the needs of different subsectors of the food industry have learned how to deal with the attending multiple problems.

In any case, the worldwide energy requirements for heating and transportation are probably mismatched to what the planet's flora can offer. The flora can sustain (and has sustained) the fauna with O_2 and the energy required for living. This sustenance includes shelter, clothing, and nutrition for humans. The flora also is the source of all fossil fuels except nuclear. However, when requested to provide the energy flows of industrial societies (vast cities, transportation), a seemingly insurmountable task arises.

ENERGY SAVINGS THROUGH ENHANCED EFFICIENCY

The heading of this brief discussion could have been energy conservation. Conservation, for most folk, is "doing without" [23]. Whereas in a broad sense efficiency includes not wasting energy, improving technology is more apt to produce energy savings. Nevertheless, wasteful ways are generally not good for individual temperance, which is worth cultivating under any circumstance, and much more so when the legacy matter is factored into one's thinking.

Enhanced efficiency implies combined cycles such as discussed in Chapter 5, improved lighting, and motors with examples given in Chapter 6, and improved space conditioning systems and insulation as discussed also in Chapter 6. A topic not covered in this text but of utmost importance is that of fenestration, which when improved, especially in older homes, can substantially reduce energy costs. Energy efficiency also covers more fuel-efficient transportation, discussed earlier, and mass transportation with suitable frequencies, comfort, and reliability as to become a credible alternative to individual transportation.

Substantial investments in energy-efficient technology lead to increased employment, as shown, for instance, in Reference 24. The development and implementation of advanced energy-efficient technology seems to ultimately be an investment in the labor force. Energy-efficient technology reduces the generating capacity, and it could be a necessary component of renewable energy. Renewable energies could likewise be perceived to enhance employment; however, this is a debated point. Although one would think that a distributed network would call for investment in labor for setup and maintenance, this is not straightforward. In countries with good prospects for renewables, the expectation seems to hold at least in some localities [25]. Other studies [26] project enhanced employment in rural areas, but not in urban ones. Finally, some economies are such that initially jobs are created, but a number are lost after renewable technology is established [27].

A CONVERGENCE TO RENEWABLES?

Although some organizations are showing the way, a transition to renewables, which seems a priority, is not necessarily easy. However it is possible, at least in some cases. A recent report [28] describes a carbon-neutral factory with a production on the order of 40,000 large trucks per year. The plant relies on photovoltaic, wind turbines, purchased hydro power, and biomass combustion to achieve carbon neutrality. The article does not report on cost or maintenance requirements of such an array of technologies but does make the point that all the technologies are proven ones. Yearly output is 14.5 GWh, with an excess on the order of 3 GWh sold to the grid.

We conclude here with two thoughts. The first concerns the challenges of renewables. Until the 20th century, mankind did not have energy systems for power and heat production anywhere comparable to those using fossil or nuclear fuels. Mankind's comfort and ability to perform work depended, directly or indirectly, on renewables. There still is enough solar and wind to sustain an enhanced world economy. Also, much more efficient ways to convert and direct energy have been devised, for our grids/ generating installations have considerable intelligence and short response times, which could feasibly result in seamless transitions from renewables to backup power when the former are absent. Harnessing renewable energies is perhaps the greatest challenge of all time facing engineers. The problems are complex, and their scale staggering, but we have much more knowledge and technology than we did before. So, whether we will earnestly pursue renewables or they are just a passing fad is really a crucial decision that each individual has to make and translate into his or her way of living.

The second thought concerns the hypothetical payback of turning to renewables. With smaller pollution levels (by using advanced technology, Chapter 5), and evenly distributed harvesting technology, the world could indeed be a better place. However, energy efficiency is a must if renewables are to become a significant component of the energy budget of industrialized nations, and that could mean higher employment. Although the technical and economic challenges are substantial, perhaps, in time, we would come to realize that of all the possible futures, one that relied considerably on renewables would be the best.

REFERENCES

1. *Perry's Chemical Engineer's Handbook*, 1997. Edited by Perry, R.H., D. W. Green, McGraw Hill, New York.
2. Ordorica-Garcia, G., P. Douglas, E. Croiset, L. Zheng. 2006. Technoeconomic evaluation of IGCC power plants for CO_2 avoidance, *Energy Conversion and Management*, Vol. 47, pp. 2250–2259.

3. Roets, P., 2006. Qualification of Sasol Fully Synthetic Jet Fuel as Commercial Aviation Turbine Fuel, TIACA Executive Conf., 4/10-4-11, 2006, China, in www.tiaca.org/content/AGM2006/Presentations/13_Piet%20Roets.pdf

4. Fairley, P., 2007. China's Coal Future, *MIT Technology Review,* http://www.technologyreview.com/Energy/18069/page3/

5. Derner, J. D., G. E Schuman, 2007. Carbon sequestration and rangelands: A synthesis of land management and precipitation effects, *Journal of Soil and Water Conservation,* Vol. 62, No.2, pp. 77–85. Retrieved June 9, 2008, from Research Library Core database. (Document ID: 1274928151)

6. Litynski, J. T., S. M. Klara, H. G. McIlvried, R. D. Srivastava, 2006. An overview of terrestrial sequestration of carbon dioxide: the United States Department of Energy's fossil energy R&D program, *Climatic Change,* Vol. 74, No. 1–3, pp. 81–95.

7. Gibbins, J., H. Chalmers, 2008. Preparing for global rollout: A "developed country first" demonstration programme for rapid CCS deployment, *Energy Policy* Vol. 36, pp. 501–507.

8. Quadrelli, R., S. Peterson, 2007. The energy–climate challenge: Recent trends in CO_2 emissions from fuel combustion, *Energy Policy,* Vol. 35 pp. 5938–5952.

9. British Petroleum, Statistical data, http://www.bp.com/liveassets/bp_internet/globalbp/globalbp_uk_english/reports_and_publications/statistical_energy_review_2007/STAGING/local_assets/downloads/pdf/table_of_world_oil_consumption_2007.pdf

10. British Petroleum, Statistical data, http://www.bp.com/liveassets/bp_internet/globalbp/globalbp_uk_english/reports_and_publications/statistical_energy_review_2007/STAGING/local_assets/downloads/pdf/table_of_proved_oil_reserves_2007.pdf

11. British Petroleum, Statistical data, http://www.bp.com/sectiongenericarticle.do?categoryId=9017911&contentId=7033476

12. British Petroleum, Statistical data, http://www.bp.com/sectiongenericarticle.do?categoryId=9017913&contentId=703344

13. Kidd, S. W., 2006. Time for a fresh look at nuclear? *Energy and Environment*, Vol. 17, No. 2, pp. 175–180.

14. World Nuclear Association. 2008. http://www.world-nuclear.org/info/reactors.html

15. World Nuclear Association. 2008. http://www.world-nuclear.org/info/inf08.html

16. World Nuclear Association. 2008. http://www.world-nuclear.org/info/inf62.html

17. World Nuclear Association. 2008. http://www.world-nuclear.org/info/inf60.html

18. Schwarzbozl P., R. Buck, C. Sugarmen, A. Ring, M. Marcos Crespo, P. Altwegg, J. Enrile, 2006. Solar gas turbine systems: Design, cost and perspectives, *Solar Energy*, Vol. 80, No. 10, pp. 1231–1240.

19. Alrobaei, H., 2008. Novel integrated gas turbine solar cogeneration power plant, *Desalination*, Vol. 220, No. 1–3, pp. 574–587.

20. Willyard, C., 2008. Switching To Switchgrass, *GeoTimes*, Vol. 53, No. 3, pp. 8–9.

21. Erbrecht T., W. Lucht, 2007. Is a substantial global bioenergy system feasible? A spatial analysis using a dynamic global vegetation model, *EOS Transactions*, American Geophysical Union, Vol. 88, No. 52, Suppl. Vol. 1–2.

22. Stillman, C., 2006. Cellulosic Ethanol: A Greener Alternative, http://www.cleanhouston.org/energy/features/ethanol2.htm

23. Deffeyes, K. S. 2005. *Beyond Oil: The View from Hubbert's Peak.* Hill and Wang, New York.

24. Quirion P., M. Hamdi-Cherif, 2007. General equilibrium impact of an energy-saving policy in the public sector, *Environ Resource Economics*, Vol. 38, pp. 245–258.

25. Moreno, B., A. Jesús López, 2008. The effect of renewable energy on employment: The case of Asturias (Spain), *Renewable and Sustainable Energy Reviews*, Vol. 2, pp. 732–751.
26. Bergmann, A., N. Hanley, R. Wright, 2006. Valuing the attributes of renewable energy investments, *Energy Policy*, Vol. 34, pp. 1004–1014.
27. Hillebrand, B., H. Buttermann, J. M. Behringer, M. Bleuel, 2006. The expansion of renewable energies and employment effects in Germany, *Energy Policy,* Vol. 34, pp. 3484–3494.
28. Appleyard, D., 2008. On-site renewable at Volvo, *Cogeneration and On-Site Power Production*, May–June, pp. 81–83.

Nomenclature

A	Area, m^2
Aw	Frontal wind turbine area, m^2
a	Polynomial coefficient (a1, a2 ...)
b	Generic intensive property, property unit/kg, or polynomial coefficient (b1, b2 ...)
B	Generic intensive property, property unit
BTM	Combustor thermal inertia, J/kg
c	Specific heat, J/(kg·K); or Polynomial coefficient (c1, c2 ...)
c_p	Constant pressure specific heat, J/(kg·K)
cp	Wind power coefficient
c_v	Constant volume specific heat, J/(kg·K)
C	Capacitance, F
C_{blg}	Building thermal capacity, J/K
$C2$	Constant, Pa·m$^{3.99}$
$C4$	Constant, Pa·m$^{3.75}$
$C2T$	Constant, K·m$^{0.99}$
C_D	Drag coefficient
CTM	Compressor thermal inertia, J/K
d	Diameter, m
D	Denotes substantial derivative or drag force, N
Dh	Hydraulic diameter, m
DI	Direct normal irradiance, W/m^2
DR	Diffuse radiation, W/m^2
DT	Clearance at the top of the piston, m
e	Specific energy, J/kg
ex	Exergy, J/kg
E	System energy, J
f	Friction factor
F	Force, N; or view factor
g	Acceleration of gravity, m/s^2
gr	Growth rate, %
H	Solar hour angle
h	Elevation, m, or specific enthalpy, J/kg
hc	Convective heat transfer coefficient, (W/m^2·K)
H	Elevation, m
HD	Household demand, kw·hr/day
I	Current, A, or moment of inertia, kg·m^2
Ir	Rotor current, A
k	Constant or thermal conductivity, W/(m·K)
$k1$	Time constant for resource multiplication, yr

KE	Kinetic energy, J
l	Rod length, m or latitude, rad
L	Length of duct, m, or inductance, He
Li	Resource lifetime
LHV	Lower heating value, J
Ln	Light flux, W/m^2
m	Mass, kg
$\dot{m}$	Mass flow rate, kg/s
M	Mass inside control volume, kg
Mfr	Massflow rate, kg/s
MF	Multiplier for resource, dimensionless
n	Exponent
N	North
p	Pressure, Pa, or momentum, kg·m/s
pe	Polytropic exponent
pr	Gas turbine pressure ratio
P	Power, W
PE	Potential energy, J
Pr	Production rate, toe/yr
PVW_{USA}	Annual average solar electromagnetic energy arriving at unit surface per day, kW·hr/(m^2·day)
Q	Heat exchanged, J
r	Logistic growth constant, 1/(toe·yr); or Crank radius, m
rpm	Revolutions per minute, 1/min
$rUSA$	USA logistic growth constant, 1/(toe·yr)
rW	World logistic growth constant, 1/(toe·yr)
R	Electrical resistance, Ω, or radius, m
Re	Proven reserve, toe
Rf	Resistance to flow, Pa/(kg/s)
Rg	Gas constant, J/kg·K
Rm	Stage mean radius, m
$RUSA$	USA proven reserves, btoe
RW	World proven reserves, btoe
Ro	Overall building thermal resistance, K/W
R_{th}	Thermal resistance, K/W
R_r	Resistance to radiative energy flow, K/W
s	Displacement, m, or specific entropy, J/(kg·K)
sm	Fixed amount of mass flowing through a control volume, kg
S	Entropy, J/K, or south
SE	Solar energy flux, br/yr·m^2
SPO	Solar power output per unit area, W/m^2
t	Time, s
T	Temperature, K
T_o	Total amount of a resource, toe
TI	Thermal input, W
TM	Thermal inertia

TTM	Turbine thermal inertia
u	Specific internal energy, J/kg
U	Internal energy, J, or amount of fuel used up to a certain point in time, toe
UUSA	Amount of resource used to date in the U.S.A., b toe
UW	Amount of resource used to date in the world, b toe
v	Specific volume, m³/kg
V	Velocity, m/s
V_r, VC, VL	Voltage drop, V, across resistor, capacitor and inductance
Vol	Volume, m³
W	Thermodynamic work, J, or Relative velocity, m/s
W2	Weibull probability
WR_{USA}	Wind resource, W/m²
x	Geometrical coordinate, m
y	Dependent variable
z	Elevation over a reference plane, m
Z	Compressibility factor

GREEK SYMBOLS

α	Angle, deg, or Weibull distribution parameter
β	Weibull distribution parameter, or Solar altitude, rad
δ	Declination angle, rad
Δ	Change
H	Hour angle, rad
ε	Emissivity (gray surface)
θ	Azimuthal direction, or angle between sun's ray and surface normal, rad
ρ	Density, kg/m³
σ	Denotes control volume, or Stefan-Boltzmann constant, $(5.67 \cdot 10^{-8}$ W/m² · K)
η	Thermal efficiency
τ	Torque
Φ	Solar azimuth, rad, or equivalence ratio
ψ	Angle between normal projection on the horizontal plane and N–S lane, rad
μ	Friction coefficient, or angle between surface normal and vertical, rad

SUBSCRIPTS

a	Air
atm	Atmosphere
batt	Battery
c	Cold
cc	Chemical engine
coal	Coal

comp	Compressor
Ca	Carnot
e	Engine
exp	Expansion
f	Fuel
fo	Initial fuel mass
g	Gas
gen	Generated
i	Input
init	Initial
inv	Inverter
h	Hot
Lost	Lost
max	Maximum
M	Mechanical
n	Denotes a time value
na	Nacelle
ng	Natural gas
o	Output
oil	Oil
pv	Photovoltaic
p	Piston
rev	Reversible
rf	Roof
Sys	System
th	Thermal
tot	Total, denoting all systems and atmosphere
turb	Turbine
ur	Uranium
va	Constant air volume
vg	Constant gas volume
vol	Volumetric
w	Wind or windmill; arrow denotes vector

UNITS

ENERGY

barrel of oil	broil	broil = 0.1364; toe = $5.73 \cdot 10^9$ J
joule	J	1
kilo J	kJ	kJ = 10^3 J
mega J	MJ	MJ = 10^6 J
giga J	GJ	GJ = 10^9 J
tera J	TJ	TJ = 10^{12} J
exa J	EJ	EJ = 10^{15} J
peta J	PJ	PJ = 10^{18} J

Ton of oil equivalent	toe	$toe = 4.2 \cdot 10^{10} \cdot J = 42$ GJ
Electrical toe	toee	$toee = 1.6 \cdot 10^{10}$ J $= 16.2$ GJ
Electrical watt-hr	we·hr	$we \cdot hr = 3.6 \cdot 10^3 \cdot J = 3.6$ kJ

TIME

second	s	
hour	hr	$hr = 3600$ s
day	day	$day = 24$ hr $= 86400$ s
year	yr	$yr = 365$ day $= 8.766$ hr $= 3.2 \cdot 10^7$s

POWER

Watt	W	J/s
kilo W	kW	$kW = 1000$ W
mega W	MW	$MW = 1000$ kW
giga W	GW	$GW = 1000$ MW

Index

Printed and bound by CPI Group (UK) Ltd, Croydon, CR0 4YY

18/10/2024

01776244-0010